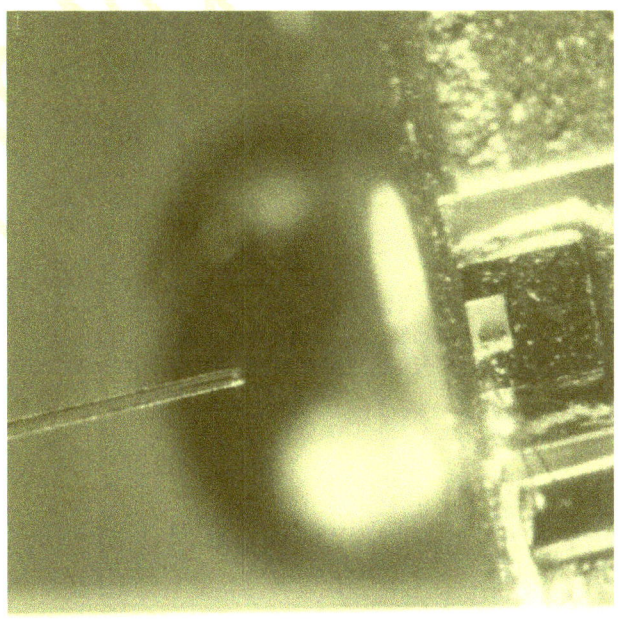

Ronan O'Dowd's Designer Books in Photonic Engineering:

Photonics Handbook Part 2

"Photonics Handbook Part 2: LEDs, Lasers, Detectors"

"Photonics Handbook Part 2: LEDs, Lasers, Detectors"

Author

Ronan O'Dowd PhD SMIEEE is Professor Emeritus Photonic Engineering at UCD Dublin, Ireland where he taught and researched Optoelectronics and Photonics for three decades until 2010. He has several breakthrough papers in topics such as tunable semiconductor lasers and optical communications, including the millennium 2001 paper proving a dense comb of 2000 wavelength channels could be transmitted in a single fibre using the same semiconductor laser (ref *IEEE Jnl.S.T. Quantum Electronics Mar 2001*). Many of his students have proceeded to successful careers in academia and the photonics industry worldwide.

By the same author: Physics Science of Action   Gill and Macmillan 1984

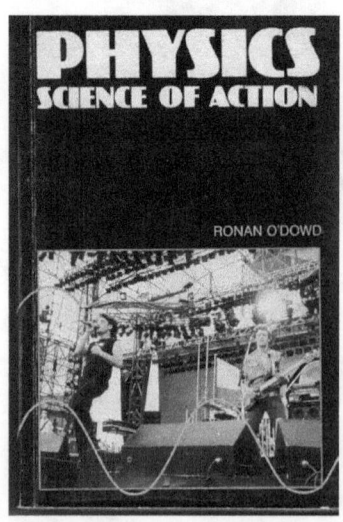

"Photonics Handbook Part 2: LEDs, Lasers, Detectors"

Tips to use this guidebook.

This series of books for photonic system designers will cover the subsystems that make up an optical communications link. These are the transmitter, fibre channel and receiver and since the fibre design sets the specifications and key criteria to be implemented at either end we tackled that in Part 1. Part 2 covers the terminals
The guide is formatted with *You Do* exercises and answers are provided. These are intended to be part of the learning process that will take the student to Engineering degree high level over what may be about a 30 hour degree module where 6 additional hours are set aside for practical work. The hardware link kit from OptoSci, where optical fibre dispersion etc can be measured over fibre reels, is very useful for laboratory explorations.

This Part 2 now covers laser, detector and system design for broadband and may constitute a further module.

COVER PICTURE: An optical fibre 1/8 mm thick faces a semiconductor laser chip seated on a metal sub-mount in the author's laboratory.

CONTENTS PART 2

Notes:

You Do exercises should be attempted especially by the self-taught student using this guide and regardless of your confidence in your answer quality. Having then read the answer provided you should again attempt it.

Diagrams are simple line-style that the student should re-draw.

There are sample examination questions at the end using standard mathematical relations.

1 Materials for Photonic Devices

The commonest semiconductor for electronics is silicon, Si, and is in group 4 (IV) of the periodic table of elements (Table 1 and Appendix 1). We will see that it can also be used, but only in restricted circumstances, for *photonics*, where light particles carry the information in place of electrons. We must therefore seek other materials that are more suited to the emission and detection of light and that is the realm of photonics. We will find that a family of semiconductor alloys consisting of two (binary), three (ternary) or four (quaternary) elements from groups 3 (III) and 5 (V) of the periodic table will satisfy our requirements adequately.

To understand these alloys we must review briefly the band theory of semiconductors as it applies to group IV silicon and group III-V alloys such as gallium arsenide GaAs and indium phosphide InP.

Table 1 Periodic Table of Elements

Group:

I	II											III	IV	V	VI	VII	VIII
1		3	4	5	6	7	8	9	10	11	12						18
1 H 1.008	2																2 He 4.0026
3 Li 6.94	4 Be 9.0122											5 B 10.81	6 C 12.011	7 N 14.007	8 O 15.999	9 F 18.998	10 Ne 20.180
11 Na 22.990	12 Mg 24.305	3	4	5	6	7	8	9	10	11	12	13 Al 26.982	14 Si 28.085	15 P 30.974	16 S 32.06	17 Cl 35.45	18 Ar 39.948
19 K 39.098	20 Ca 40.078	21 Sc 44.956	22 Ti 47.867	23 V 50.942	24 Cr 51.996	25 Mn 54.938	26 Fe 55.845	27 Co 58.933	28 Ni 58.693	29 Cu 63.546	30 Zn 65.38	31 Ga 69.723	32 Ge 72.63	33 As 74.922	34 Se 78.96	35 Br 79.904	36 Kr 83.798
37 Rb 85.468	38 Sr 87.62	39 Y 88.906	40 Zr 91.224	41 Nb 92.906	42 Mo 95.96	43 Tc (98)	44 Ru 101.07	45 Rh 102.91	46 Pd 106.42	47 Ag 107.87	48 Cd 112.41	49 In 114.82	50 Sn 118.71	51 Sb 121.76	52 Te 127.60	53 I 126.90	54 Xe 131.29
55 Cs 132.91	56 Ba 137.33	57-71 *	72 Hf 178.49	73 Ta 180.95	74 W 183.84	75 Re 186.21	76 Os 190.23	77 Ir 192.22	78 Pt 195.08	79 Au 196.97	80 Hg 200.59	81 Tl 204.38	82 Pb 207.2	83 Bi 208.98	84 Po (209)	85 At (210)	86 Rn (222)
87 Fr (223)	88 Ra (226)	89-103 #	104 Rf (265)	105 Db (268)	106 Sg (271)	107 Bh (270)	108 Hs (277)	109 Mt (276)	110 Ds (281)	111 Rg (280)	112 Cn (285)	113 Uut (284)	114 Uuq (289)	115 Uup (288)	116 Uuh (293)	117 Uus (294)	118 Uuo (294)

* Lanthanide series														
57 La 138.91	58 Ce 140.12	59 Pr 140.91	60 Nd 144.24	61 Pm (145)	62 Sm 150.36	63 Eu 151.96	64 Gd 157.25	65 Tb 158.93	66 Dy 162.50	67 Ho 164.93	68 Er 167.26	69 Tm 168.93	70 Yb 173.05	71 Lu 174.97

# Actinide series														
89 Ac (227)	90 Th 232.04	91 Pa 231.04	92 U 238.03	93 Np (237)	94 Pu (244)	95 Am (243)	96 Cm (247)	97 Bk (247)	98 Cf (251)	99 Es (252)	100 Fm (257)	101 Md (258)	102 No (259)	103 Lr (262)

You do...

Ex 1: Elements for photonics.

Locate the elements silicon Si, gallium Ga, arsenic As, indium In, and phosphorous P by period and group in the Periodic Table of elements Table 1 / Appendix 1. Draw a sub-set of the main table three groups wide and two periods deep containing all of these elements.

Answer Ex 1 at end handbook and in Appendix 1.

 o End of exercise

The energy band diagrams of Si and GaAs, both being semiconductors, consist of a conduction band separated by a forbidden gap E_g from the valence band, Figure 1.

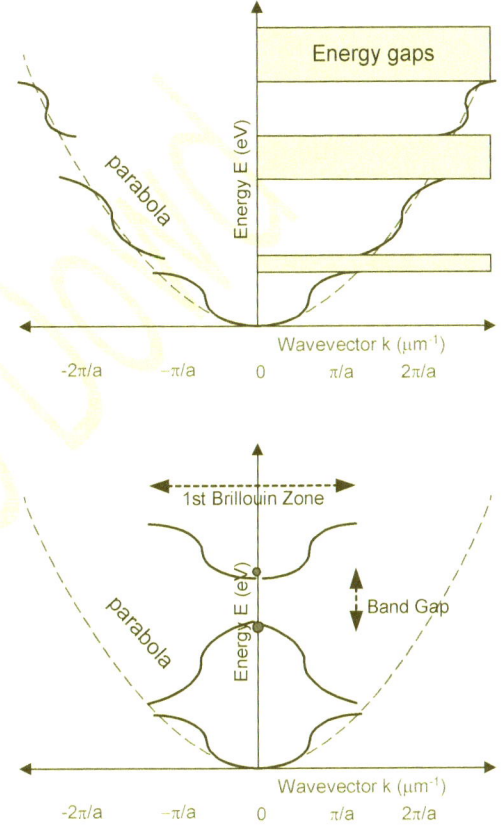

Figure 1. Band diagrams show E (energy) versus k (wave-vector) for photonic semiconductors. The top two bands are then isolated and segments translated to the +/- π/a range of k to produce the reduced Brillouin zone representation.

These diagrams are derived in semiconductor coursework by applying the Schroedinger equation in the context of the periodic potential V(x) experienced by the electron waves in a crystal of periodicity a. In the one dimensional or 1D case $V(x) = V(x+a) = V(x+2a) = $ ...etc. The outcome of that analysis is the allowed energies showing discontinuities along the electronic wave vector axis k wherever $k = +/- n\pi/a$ (Figure 1 upper). The top two bands are then isolated and segments translated to the $+/- \pi/a$ range of k to produce the reduced *Brillouin zone* representation (Figure 1 lower).

The upper two levels alone are of interest to us here since lower bands require X-ray energies to access them. Visible and infra-red or IR energy light can access the two upper bands, called conduction and valence, that are separated by a forbidden gap E_g. That gap is about 1 electron volt, 1 eV, depending on the material.

There is a crucial distinction between the E-k diagram for Si, Group IV, and that for GaAs or the other III-V alloys. With GaAs (Figure 2 upper) the bottom of the conduction band resides directly above the top of the valence band so it is called a *direct bandgap semiconductor*. For Si however this is not the case (Figure 2 lower) so it is called an *indirect bandgap semiconductor.* This means that along with an energy gap E_g between these positions on the diagram there is also a horizontal gap along the k-axis where wave-vector k is related to and a measure of the momentum p for the electron.

Recall: electron wave-vector k = $2\pi/\lambda$ while momentum p = h/λ. Therefore p is proportional to k and the E-k band diagram effectively plots electron energy versus momentum.

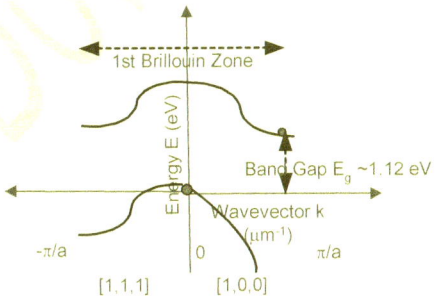

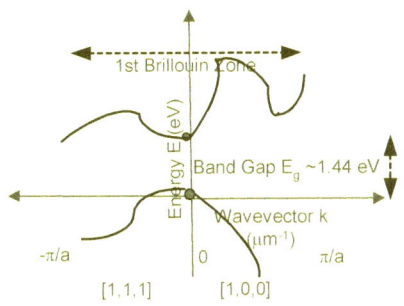

Upper represents Si indirect band gap 1.12 eV; lower GaAs direct bandgap 1.44 eV

Figure 2. Band diagrams for GaAs and Si showing added detail with crystal directions [1,1,1] at left of E-axis and [1,0,0] at right of E-axis. (This is done for data compression, otherwise left is just a reflection of right).

An electron in gallium arsenide that recombines with a hole must transit vertically releasing energy equal to E_g and that can be in the form of a photon of light. (See Ex 2). For this to happen with silicon however there must be a momentum shift as well in order to move horizontally along the k-axis from bottom of conduction to top of valence bands (dots in Figure 2). To obey the law of conservation of momentum a third particle must participate that gives or takes away momentum. There is only a slight probability that a third particle exists nearby at this instant so the likelihood of a photon being created in this way is very small. *Injection luminescence,* the emission of light by inputting current, is not expected from silicon or indeed found except in novel, artificial forms of that element.

You do...

Ex 2: Wavelengths and energy gaps.

Calculate the wavelength associated with a 1 eV energy gap and also for Si and GaAs where it is 1.12 eV and 1.44 eV respectively.

Answer Ex 2 at end handbook.

○ End of exercise

The 3D solution to the Schroedinger equation is more complicated but similar diagrams to 1D apply corresponding to different directions in the crystal. The 1st Brillouin zone has shape dependent on crystal structure. Boundaries are still close to π/a where a is defined by the unit cell dimension. In real crystals the maximum of the valence band does not always occur at the same k value as the minimum of the conduction band. In an indirect bandgap semiconductor they do not while in a direct bandgap semiconductor they do coincide.

For this reason Si is not used in photonics for light emitting diodes, *LEDs*. Si may be used to detect light however as long as the light wavelength is in the visible or near IR and shorter than 1100 nm or 1.1 μm, corresponding to the bandgap of silicon (Exercise 2).

III-V alloys are the prime choice for luminescent semiconductors in the form of LEDs. These may be binary, ternary or quaternary alloy mixes depending on the light wavelength we wish to create. See Table 2 and

Figure 6 below. They will detect at those IR wavelengths also so are suitable for photodiodes too.

Other materials that are indirect bandgap can be used to create photodiode detectors to suit particular wavelengths such as the 1800 nm region where germanium, Ge, has a bandgap. Broadband optical fibre operates in the 1550 nm window so that ternaries and quaternaries are the choice for both transmitter TX and receiver RX.

Injection Luminescence

The more detailed band diagrams for Si and GaAs are shown in Figure 2 above. GaAs is grown in a similar fashion to Si by *chemical vapour deposition, CVD* using extremely pure reactant gases. The carrier gas valves are programmed to allow input of elemental dopants at precise concentrations to produce layers that are p-type and n-type. By these means a wafer of pn diodes results. The wafer is then cut into chips that are packaged to give a light emitting diode or LED.

When current is injected into a positively-biased pn-diode the movement of electrons and holes is well known. With heavy doping and under forward bias minority carriers are injected from both sides Figure 3. The excess electron concentration at distance x from the junction in the p material falls exponentially with x.

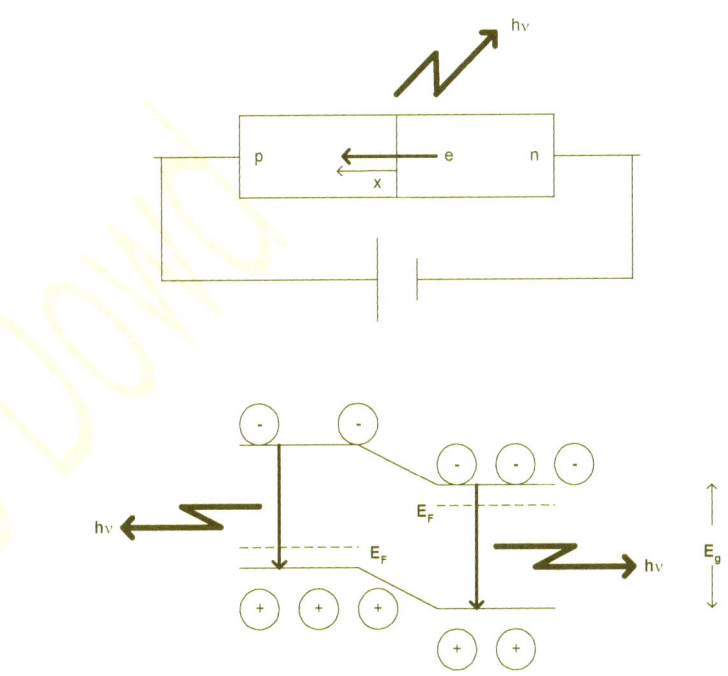

Figure 3. GaAs LED with forward bias showing carrier injection of electrons in the conduction and holes in the valence bands. Recombination after a penetration distance x produces IR light photons $h\nu$ corresponding to the energy gap E_g.

Quantum efficiency is the rate of emission of photons divided by the rate of supply of electrons by injected current. The units are therefore watts of light per amp of current, WA^{-1}.

If *radiative recombination* does occur a photon is produced with wavelength λ_g given by Plank's equation:

$$\lambda_g = hc/E_g$$

Thermal energy ensures electrons in the conduction band have average energy $kT/2$ above the bottom of the band so that λ can be shorter than λ_g due to an increased jump. Meanwhile much recombination involves dopant energy levels within the gap so that λ may more likely be longer than λ_g producing a spread of "colours", the emission spectrum Figure 4. The above may be avoided when a *phonon* is available (a particle carrying the energy of a crystal vibration, essentially becoming heat) and then non-radiative recombination reduces the quantum efficiency.

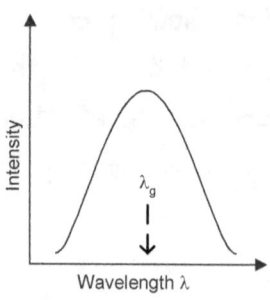

Figure 4. Emission spectrum for GaAs LED centred on λ_g.

You Do...

Ex 3: Optical power from LED.

A forward current of 30 mA is injected into a GaAs LED with quantum efficiency 95%. What output light power results? Use data from Ex 2.

Answer to Ex 3 at end handbook.

Interband Transitions

Transitions must conserve total wave-vector of the system. For a photon $k = 2\pi/\lambda$ while for an electron the range is $-\pi/a$ to $+\pi/a$ or $2\pi/a$ as shown in the Brillouin-zone diagram Figure 1. For 500 nm radiation (a visible LED) and typical lattice spacing 1 angstrom or 0.1 nm we find $2\pi/\lambda$ is thousands of times smaller than π/a. (Verify as an Exercise). Hence if only a photon and electron are involved then the transition is between states with virtually the same electron wave-vector (Figure 5a). If a phonon is also involved then a non-vertical transition may be possible (Figure 5b). In the latter case the phonon may contribute its energy E_p by annihilation or deduct E_p by being created as a crystal vibration from the process:

$$hc/\lambda = E_g +/- E_p$$

Phonon energies are typically only 0.01 eV so that the light wavelength is approximately equal to λ_g while momentum is still conserved. However since three particles must be involved in the same place at the same time this transition probability is very low.

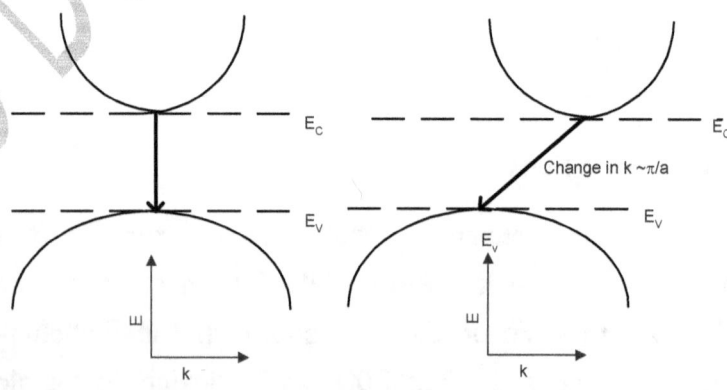

Figure 5. Direct and indirect transitions.

Interband *transition rate* r may be written:

$$r= Bnp$$

Here the participant carrier concentrations are n electrons and p holes while the coefficient B is assigned to each material. Note from Table 3 that for direct bandgap semiconductors B is approximately 10^5 times greater than indirect bandgap materials like Si. Unless other mechanisms are present, therefore, the latter, like silicon, are not suitable for LEDs.

TABLE 3   Common photonic materials

Material	Bandgap	E_g (eV)	B ($m^3 s^{-1}$)	λ_g (nm)
Si	indirect	1.12	1.79×10^{-21}	1106
Ge	indirect	o.67	5.25×10^{-20}	1880
GaP	indirect	2.26	5.37×10^{-20}	549
GaAs	direct	1.44	7.21×10^{-16}	861
InP	direct	1.35	1.26×10^{-15}	918
CdTe	direct	1.50	library task	826

Alloy Map for InGaAsP

Various mixes of the four elements indium In, gallium Ga, arsenic As, and phosphorous P produce injection luminescence at different wavelengths because the energy gap depends on the alloy chosen. The mole fraction for In is designated x so that for Ga must be

(1-x) as one substitutes for the other in the crystal structure. Equally if As has mole fraction y then P must be at (1-y). The alloy is designated $In_xGa_{1-x}As_yP_{1-y}$ and the lattice parameter or atomic spacing in the crystal depends on x and y. Since the energy gap E_g is dependent on the lattice parameter a we may draw an alloy map that summarises the important photonic information, Figure 6.

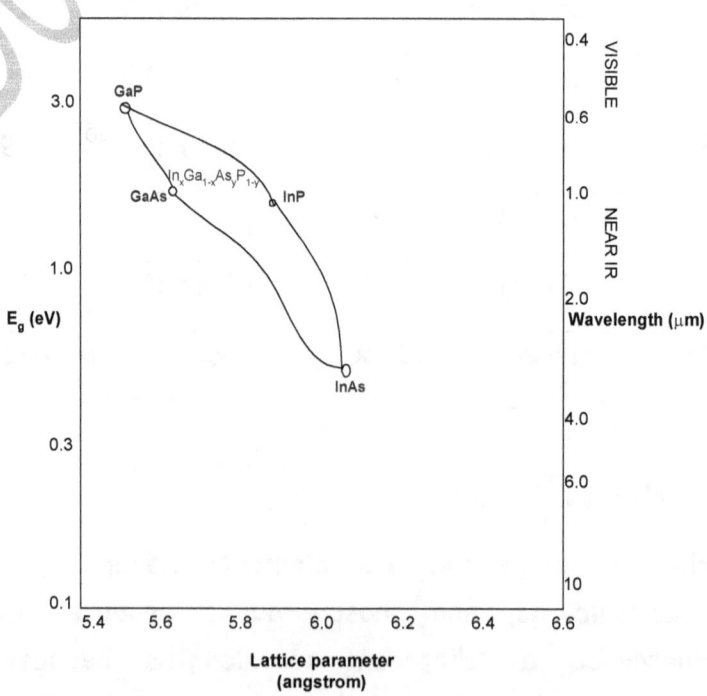

Figure 6. Alloy map for $In_xGa_{1-x}As_yP_{1-y}$ with ternary compounds along the perimeter and quaternaries inside.

The vertical axis can be both E_g and λ as $E_g=hc/\lambda$ and this is fundamental information to selecting the most suitable material for a given diode design. It is evident that we can select x and y to create visible or infra red LEDs and lasers by this means called *band-gap engineering*. A selection of relevant data for the binaries, these reside at corners of the map, is given in Table 4.

Table 4.   Physical properties of photonic materials

	Units	InP	GaAs	Si
Lattice parameter	angstrom 10^{-10} m	5.9	5.6	Library exercise
Band gap	eV	1.35	1.43	
Optical transition		Direct	Direct	
Electron mobility	cm^2/Vs	4500	8500	
Hole mobility	cm^2/Vs	150	420	

Exercise: Find in the library and fill into table 4 the relevant data for Si. Calculate and fill a row for λ_g also.

LED Construction and Efficiency

The map is used and a selected alloy is grown by liquid phase epitaxy LPE or vapour phase deposition VPD onto a wafer substrate and then diced into chips which are metallised for electrical contact, Figure 7.

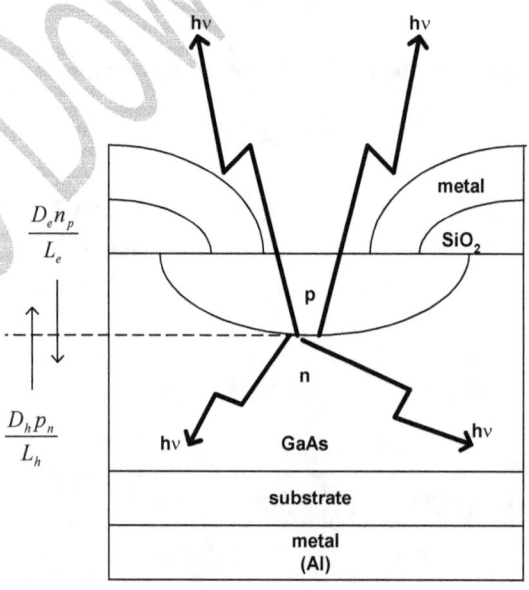

Figure 7. Surface emitting GaAs LED for 850 nm wavelength in the near-IR with insulator SiO_2 and metallisation contacts.

This is a simple surface emitting LED whose active region, where the photons are created, is designed for 850 nm light in the near-IR. A disc window is etched into the surface layer to permit exit of the radiation through the silicate insulator SiO_2 that covers the upper p-layer of the pn structure. Measures of injected carriers $D_e n_p/L_e$ and $D_h p_n/L_h$ are shown that contribute to the diode current. D is the relevant diffusion coefficient and L the diffusion length for electrons or holes while n_p and p_n are electron and hole carrier concentrations in the p and n materials respectively.

Internal Quantum Efficiency

Radiative recombination must mainly occur on the side of the pn junction nearest the surface in order to lessen re-absorption. Hence we ensure that most forward current is due to carriers injected up into the surface p-layer of this "n-side down" device. The fraction q_e carried by electrons injected into the p-side is the internal quantum efficiency as it is these that contribute to emitted power:

$$q_e = (D_e n_p/L_e)/[(D_e n_p/L_e) + (D_h p_n/L_h)]$$

Dividing by the upper bracket gives

$$q_e = [1 + (D_h L_e p_n/D_e L_h n_p)]^{-1}$$

Now use the Einstein relation for electron or hole D with mobility μ namely $D_{e,h} = (kT/e)\mu_{e,h}$ and also the basic semiconductor equation

$$n_p p_p = n_n p_n = n_i^2$$

The result becomes

$$q_e = [1 + (\mu_h p_p L_e / \mu_e n_n L_h)]^{-1}$$

In III-V compounds electron mobility μ_e is largely dominant over holes μ_h while L_e and L_h are similar in value and cancel. Now assuming we design a n^+p device where the doping n_n greatly supersedes p_p the outcome is that the inner bracket is tiny. This means q_e is almost unity or quantum efficiency approaches 100%. Observe that all this derives in the case of an n-side down device.

Exercise: If it were to be p-side down you should follow the argument through again to find that q_h can be made close to 100% by using a p^+n doping regime.

Exercise: Re-draw Figure 7 showing the depletion layer (from semiconductor theory) each side of the junction where the doping is n^+p type. What can you infer from this about efficiency in terms of location for recombination and resultant injection luminescence? In other words where are the photons created that improves efficiency?

External Quantum Efficiency

Whatever about internal value the *external quantum efficiency* for a basic surface emitting LED is much lower than 100% because of the difficulty of extracting all the radiation. Only light within a cone defined by total internal reflection TIR and the critical angle C can escape, Figure 8. The high refractive index n_1 for III-V compounds mitigates against extraction due to a small C.

By Snell's law

$$n_1 \sin i = n_2 \sin r \text{ and } r = 90 \text{ deg before TIR}$$

Hence $\qquad C = \sin^{-1}(n_2/n_1) = n_2/n_1 \qquad$ for small angles

Since n_2 is for air and n_1 is large the C value is small.

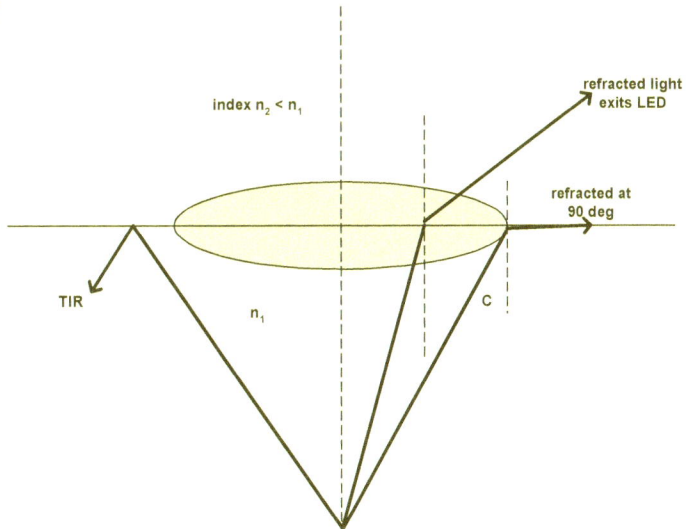

Figure 8. A cone of light defined by C escapes from the LED through the surface window; the rest suffers TIR.

Furthermore the fraction F that escapes is determined by the lit disc area relative to the surface area of a sphere filled uniformly with radiation since the emission is spontaneous and random in direction, Figure 9.

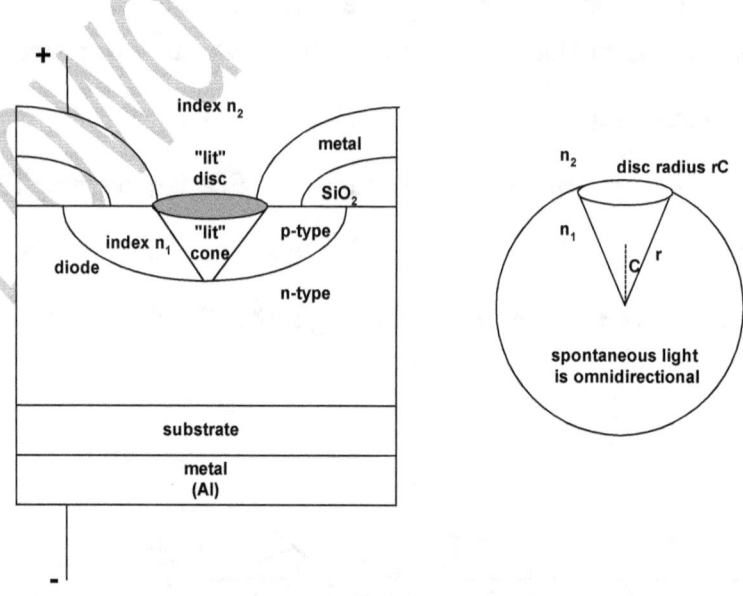

Figure 9. Cone of light that escapes through a lit disc of radius rC is determined by the critical angle C for GaAs.

Sphere surface $A_s = 4\pi R^2$ and lit disc area A_d is $\pi(rC)^2$ so that

$$A_d/A_s = C^2/4 = \tfrac{1}{4}(n_2/n_1)^2$$

Also we apply Fresnel's equation for transmission fraction at a boundary near normal incidence:

$$T = [1 - (n_1 - n_2)^2 / (n_1 + n_2)^2]$$

The total fraction extracted becomes:

$$F = \tfrac{1}{4}(n_2/n_1)^2 [1 - (n_1 - n_2)^2 / (n_1 + n_2)^2]$$

This value is the external quantum efficiency for the surface emitter and is quite small, only a percent or two, as the following exercise shows.

You Do...

Ex 4. Surface emitter efficiency.

Show the critical angle for the GaAs surface emitter with refractive index 3.6 is small and the resultant external quantum efficiency is order 1%.

Answer Ex 4 at end handbook.

○

There is an immediate improvement in the external quantum efficiency if the upper p-layer surface were curved as a sphere to ensure rays strike at near 90 deg, Figure 10a. This classical optics trick is an expensive solution as it would require etching each chip to a dome shape. The alternative is shown in Figure 10b where a dome cap is added to the surface by encapsulating the chip in transparent plastic. Now the benefit of the spherical geometry is achieved more cheaply and additionally the refractive index of plastic around 1.5 helps greatly by altering the critical angle relative to an air interface. In the *Burrus-type LED* a pit is etched at the exit window to further improve extraction of light.

Exercise: Calculate the critical angles C for GaAs to plastic and for plastic to air.

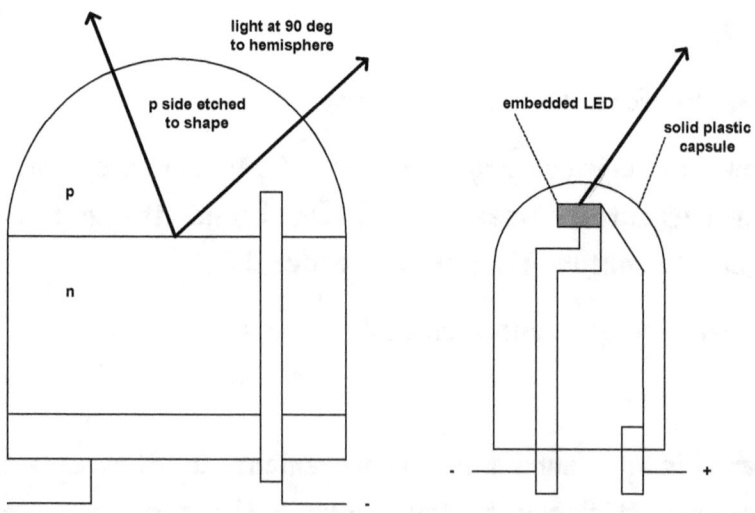

Figure 10. Hemisphere-shaped surface improves light extraction.

Edge Emitting LED, ELED

The deficiencies of the surface emitting LED are clear from the discussion so far. The *irradiance* is light power surface density launched into the forward direction, units $Wcm^{-2}sr^{-1}$. It is determined by injection current density, internal quantum efficiency, thickness of the recombination region and internal re-absorption since the GaAs energy gap also suits photon absorption. This combination of factors points to the edge-emitter design, Figure 11, as a great improvement, in fact x5 times better than a Burrus-type surface emitter. The modulation speed for the ELED is also superior. The irradiance for a Burrus-type surface emitter is about 200 $Wcm^{-2}sr^{-1}$ while for the ELED it is ~1000 $Wcm^{-2}sr^{-1}$ where the units refer to light density on unit area and with divergence contained within a solid angle of unit steradian. This specification is important for *coupling* the radiation into optical fibres as the more "contained" the light is the better and the system loss budget will be improved. It is also important for focussing light to a fine spot on a CD-ROM disc for *read/write* operation in the storage process.

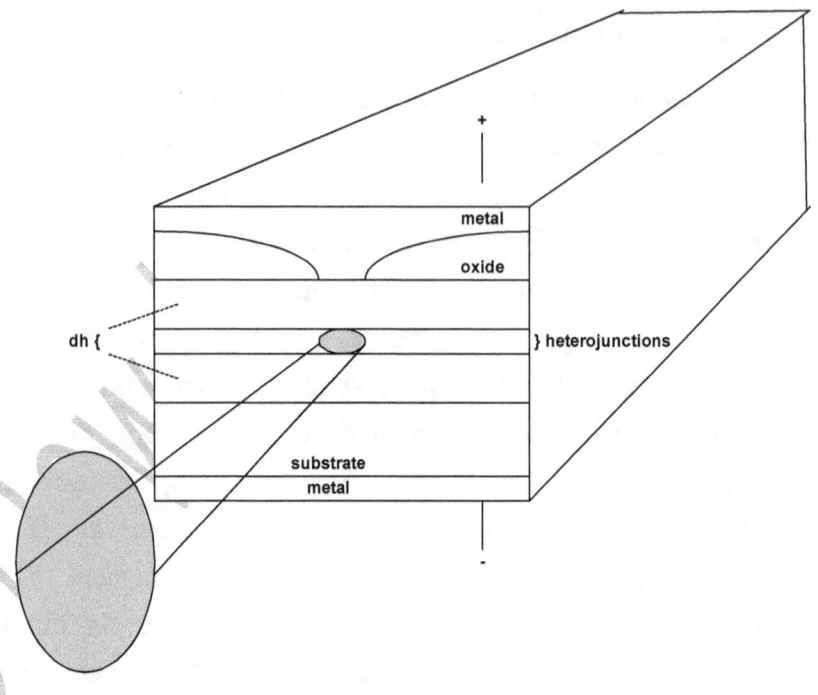

Figure 11. ELED comprised of ternary material AlGaAs with stripe contact and double heterojunction dh.

The AlGaAs ELED structure shown in Figure 11 comprises a substrate carrying a dh or double heterojunction wherein the pn diode processes take place and the resultant light is guided transversely down to the chip edge somewhat like in an optical fibre. This is called an in-plane design, referring to the junction plane. A very bright lit up window results at each end and the exiting light carries forward with it the "containment" features imposed by the semiconductor guide, namely improved irradiance. Above the dh layers there is an insulator oxide with a stripe-shaped gap and a metal layer contact on top that forces injected current into a descending sheet. That in turn provides high current density within the active region. The light arriving at the exit facets of the chip is now confined to a bright spot plotted as in Figure 12.

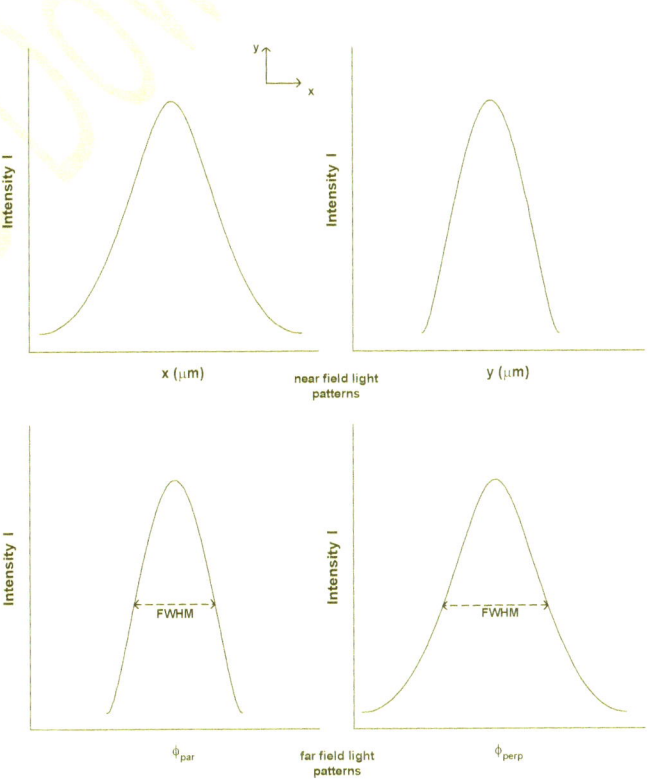

Figure 12. Near field (a) and angular far field (b) light patterns for a stripe contact dh ELED.

The *near field* pattern is one plotted as light intensity in Wcm^{-2} against distance in μm across the chip facet from one side to the other. The *far field* plots intensity against angle in degrees from the chip axis for directions parallel ϕ_{par} and perpendicular ϕ_{perp} to the junction. The latter is important for focussing onto the axis of an optical fibre. As the radiation now derives from a slit shaped lit window it diffracts as light would through a narrow slit, well known for classical optics. So the pattern at 90 deg to the plane of the slit, and equally the dh layers, spreads out more than that in the orthogonal direction as shown in Figure 12. The intensity falls off with angle to half its central value at angle 15 deg parallel to the junction but 25 deg perpendicular to the junction. Alternatively we say that the *full-width at half maximum* FWHM is 30 deg parallel and 50 deg perpendicular to the junction.

As for the near field we can infer that the lit spot at the exit facet will be oval, wider in the sideways than vertical direction. Typically this oval is only 1 μm vertical (y) by 10 μm sideways (x) and therefore a much improved tiny spot for focussing into fibre with 50 μm core or onto a storage disc. In the latter case a shorter wavelength gives even smaller *focussed spot-size* for high density data storage. We should therefore prefer UV or visible blue light to IR for read/write operations and we must select different material band gap accordingly.

Double Heterojunction or Heterotructure

A common silicon diode consists of p-type and n-type doped Si forming a homojunction of the same material. For ELEDs we devise a *heterojunction* where the p and n sides consist of different alloys by varying the mole fraction x in $Al_xGa_{1-x}As$ for example or x and y in $In_xGa_{1-x}As_yP_{1-y}$ if a quaternary semiconductor is required. Here the mole fractions x and y determine the band gap E_g and resultant wavelength λ. The heterojunction structure allows us to tailor both the energy gap and refractive index variation across the layers thereby controlling electron and light confinement at the same time. A typical dh device is depicted in Figure 13 with five different alloys.

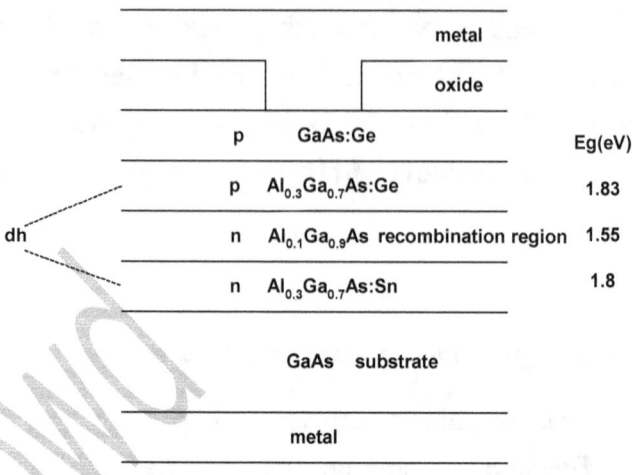

Figure 13. ELED dh structure using $Al_xGa_{1-x}As$.

For the active layer n-doped $Al_{0.1}Ga_{0.9}As$ is chosen with a band gap of 1.55 eV and recombination takes place here producing light at wavelength 800 nm. (Exercise: Verify this by calculation). This is guided by TIR until it reaches the end facets since the refractive index in $Al_{0.3}Ga_{0.7}As$ layers either side is lower. But since the energy gap 1.8 eV either side is greater than 1.55 eV the carriers are simultaneously confined to the lit region. Outside the dh structure there is the basic substrate layer below and a Ge doped p-type layer above for good ohmic contact on the p-side. A stripe at the metal-to-oxide contact ensures current *confinement* to a descending sheet. In all then there are three types of confinement, light, carriers and current, while re-absorption is also diminished by having a larger band gap than the photons on either side of the active layer. The double heterostructure greatly enhances the efficiency while creating a waveguide in the lit region.

Response Time for LED

The LED has a diffusion capacitance due to storage of carriers within a diffusion length or so of the junction. When applied voltage drops due to modulation by data these must diffuse away and then recombine to enable a new equilibrium to be established. For good frequency response the minority carrier lifetime τ must be small.

$$\tau = (Bp)^{-1}$$

Here p is majority concentration and B is the constant previously defined by r = Bnp which combined with r = n/τ gives the equation above. It is evident from this equation that τ may be reduced by heavy doping p thereby giving higher bandwidth. However near the solubility limit for acceptor impurities in GaAs non-radiative recombination centres are formed. With Ge in GaAs for example the external quantum efficiency drops above 10^{24} atoms per cubic metre. At that point using B from Table 3 above gives:

$$\tau = [(7\times10^{-16})10^{24}]^{-1} = 1.4\times10^{-9} \text{ s}$$

This result, 1.4 nanoseconds, suggests bandwidth around 700 MHz (the inverse) but in practice that is further reduced several times by stray effects from bond wires, packaging etc.

The alternative to high doping we will now show is heavy forward injection current density J with narrow active region thickness t. If Δp is injected carrier concentration and it greatly exceeds the equilibrium density p then

$$\tau = (B\Delta p)^{-1}$$

Now in equilibrium the number of recombinations each second in distance t is J/e (carriers per second per square μm) so that in unit volume the rate is J/te (carriers per second per cubic μm). But Δp recombine in 1 μm^3 in time τ so this rate is also $\Delta p/\tau$ and hence

$$J/te = \Delta p/\tau$$

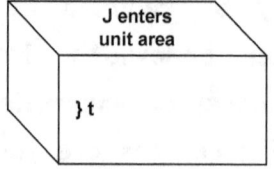

This produces $\quad \Delta p = J\tau/te$

Inserting this Δp into the above equation for τ gives:

$$\tau = (et/JB)^{1/2}$$

In order to reduce this τ value and achieve higher modulation speed we must reduce thickness t and raise injection current density J. Note however that since τ is now current dependent, or non-linear, this produces signal distortion as the modulation rises and falls.

Basic LED Drive Circuits

Two very simple schemes to derive modulated light from a LED are illustrated in Figure 14. For the first primitive circuit the LED specifications from the supplier data sheet are V_{ON} and I_{MAX} so a limiting resistor must be selected so that R_{LIM} maintains current within allowed range. When power supply voltage is set at V clearly R_{LIM} is $(V-V_{ON})/I_{MAX}$ and the switch may be manual or used to represent the modulation.

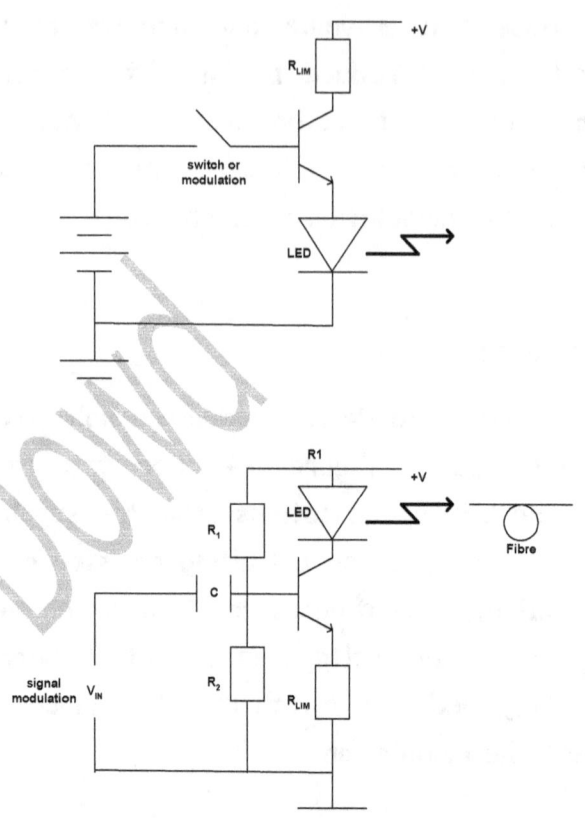

Figure 14. Primitive LED modulation circuits.

In the second scheme V_{IN} modulates diode current while R_1 and R_2 bias the transistor to half LED maximum. Both transistor and LED are in their linear regimes. The rear facet light of the LED could additionally face a photodiode that provides feedback on the level of optical power.

LED Modulation Bandwidth

The LED is normally deployed in a system comprising transmitter TX, receiver RX and optical fibre channel. Electrical injection I_{IN} supplies carriers so that photons are produced at a proportional rate and the optical power P_{OPT} is dependent on drive current I_{IN}. The electrical power however is proportional to I_{IN}^2 as shown in Figure 15.

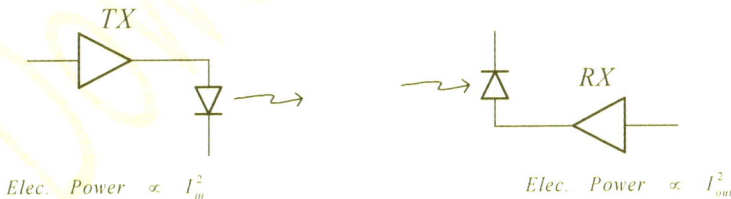

Elec. Power $\propto I_{in}^2$ *Elec. Power* $\propto I_{out}^2$

Figure 15. The optical fibre resides between two other sub-systems TX and RX.

The received light produces electrons by photon absorption so the electrical current I_{OUT} varies with the diminished P_{OPT} but the electrical power at RX depends as ever on I_{OUT}^2. The diminution of received P_{OPT} derives from the losses in the fibre and connectors.

Electrical regime dc-to-hf (high frequency) 3dB bandwidth bw is defined by that modulation frequency for which $I_{OUT}= I_{IN}/(2)^{\frac{1}{2}}$

Optical regime bw is defined by the modulation frequency for which RX optical power falls by $\frac{1}{2}$. This implies IOUT has fallen also so it corresponds to an electrical power attenuation of 6 dB   (factor 4).

If the response has Gaussian shape as expected then:

$$\text{Optical bw} = 2^{\frac{1}{2}} \text{ Electrical bw}$$

At the LED there is a time response to modulation of the drive current and this τ is combined with signal frequency ω or $2\pi f$ according to the Gaussian response function:

$$P(\omega)/P_{dc} = [1+(\omega\tau)^2]^{-1/2}$$

$P(\omega)$ is the mean modulated optical power with constant peak-to-peak current.

Pdc is optical power for the same dc drive e.g. 50 mA dc versus +/- 25 mA rf (radio frequency).

This means that optical power falls off relative to the output for a dc drive of the same amplitude whenever faster data modulation ω is deployed and equally with longer response time τ.

For edge emitters injected carrier density is the dominant factor for τ. Bimolecular recombination processes involving multiple carriers can reduce τ as do recombination at crystal defects:

$$N_2/\tau = AN + BN^2 + CN^3 = BN^2 \text{ approx.}$$

$$= \text{defects+radiative+Auger(phonon processes)}$$

B is given in Table above and is $\sim 10^{-16} \text{m}^3\text{s}^{-1}$

Thin active-layer AlGaAs LEDs have electrical bw ~250 MHz but quaternaries have corresponding bw that is 2 to 3 times higher due to shorter τ where the recombination coefficient B is large and there are more non-radiative recombination centres.

The FDDI (Fibre data distributed interface) LAN standard (Local area network) at 1.3 μm window is above 200 Mbit/s so we must select quaternary LEDs for that.

Example: LED modulation response.

A LED has response time 5 ns and produces light output 300 μW for a given dc drive. If modulated at (a) 20 MHz and (b) 100 MHz what is the optical output? Calculate also the half power frequencies.

Answer. τ = 5 ns P_{OUT} = 300 μW $\quad$ f = 20 MHz

$$P(20 \text{ MHz}) = P_{dc}/[1+(2\pi 20\times 10^6 \times 5\times 10^{-9})]^{1/2}$$

$$= 254.2 \ \mu\text{W}$$

Repeat using f = 100MHz: P(100 MHz) = 90.9 μW

For half power: $\quad 1/[1+(\omega\tau)^2]^{1/2} = \frac{1}{2}$ and f = $\omega/2\pi$

Hence $f_{optical}$ = 55.1 MHz $f_{electrical}$ = $55.1/(2)^{1/2}$ = 39 MHz

3   Lasers

To operate a LED as a laser diode design we must visit some basics concerning *light amplification by stimulated emission of radiation* or l-a-s-e-r. Diode drive current populates the upper energy level with electrons and for laser action we require *population inversion* where the upper number N_2 exceeds that for the lower level N_1 as depicted in Figure 16.

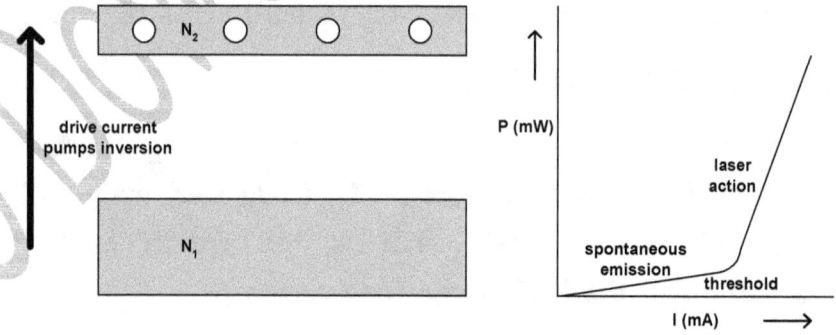

Figure 16. Population inversion in the semiconductor.

Next we expect some *stimulated emission* when a photon passes with energy corresponding to the gap E_g in the excited material. At this point we have a *semiconductor optical amplifier* or SOA because for each input photon we can get two out as recombination takes place, Figure 17.

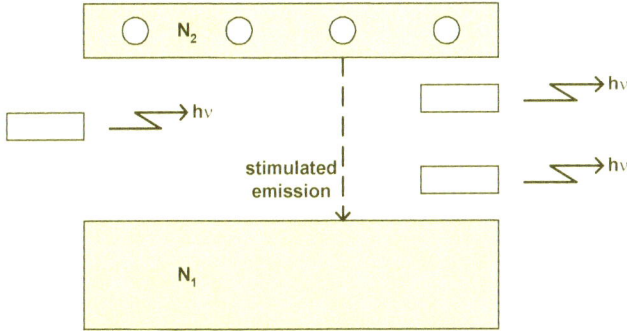

Figure 17. Stimulated emission produces a photon of equal frequency and energy to the injected photon.

To convert an amplifier to an oscillator it is known from electronics that feedback is required but in this case it should be *optical feedback*. A mirror at one or both ends will provide this. In our case the active material is a semiconductor so the partial mirror can be the polished or cleaved end facets of the chip. The partial reflectivity is found by Fresnel's equation and this two-facet mirror arrangement is called a FP or *Fabry-Perot cavity*. As the drive current is raised above a threshold the population inverts, stimulated exceeds spontaneous LED emission and the device starts to lase. This onset is accompanied by a rapid rise in optical power, Figure 16, a sharp narrowing of the optical spectrum around the energy gap frequency, Figure 18, and the appearance of coherent properties in the output light. *Cavity modes* are said to oscillate and these appear in the spectrum at higher resolution as lines for optical frequencies that are supported by the cavity length.

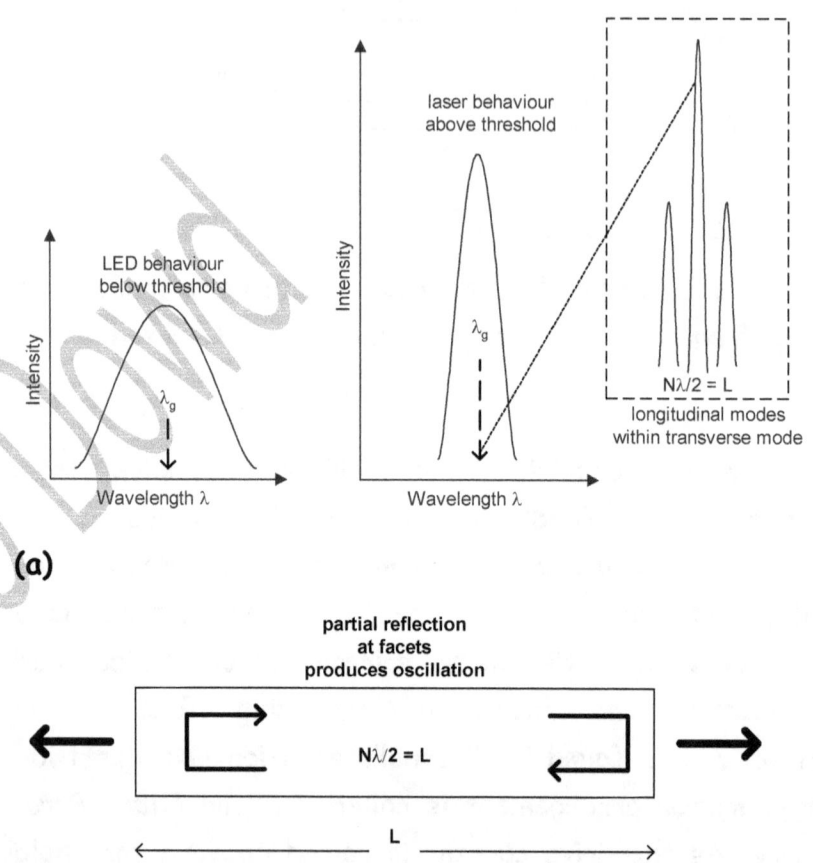

(a)

(b)

Figure 18. (a) Laser spectrum narrows sharply;(b) Chip with feedback facets that create an optical oscillator.

Einstein Equations

The three processes that may occur when light interacts with the material medium are photon absorption, spontaneous and stimulated emission. These were combined by Einstein using weighting coefficients A and B for each process and the populations for the lower and upper states N_1 and N_2.

The absorption may occur only when a photon field is present so it is stimulated and assigned B_{12}. Spontaneous emission is given A_{21} so that τ for that rate equals $1/A_{21}$.while stimulated emission is weighted B_{21}. Actual values depend on the properties of particular atoms and states according to molecular orbital theory.

In thermal equilibrium the populations of atoms were related by Boltzmann:

$$N_1/N_2 = \exp[(E_2-E_1)/kT] = \exp(\Delta E/kT)$$

Electromagnetic wave energy per unit volume per unit frequency span about frequency f is from Plank:

$\rho(f) = 8\pi hf^3/c^3[\exp(hf/kT)-1]^{-1}$ black-body radiation law

Stimulated absorption probability in interval δt is coefficient x density photons x time: $\quad B_{21}\rho(f_{21})\delta t$

Stimulated emission probability in interval δt: $B_{21}\rho(f_{21})\delta t$

Spontaneous emission probability in interval δt: $A_{21}\delta t$

Rate = probability (per unit time) x number atoms available (per unit volume). At balance excitations and relaxations cancel:

$$A_{21}N_2 + B_{21}\rho(f_{21})N_2 = B_{12}\rho(f_{21})N_1$$

Hence $\rho(f_{21}) = A_{21}/B_{21}/[(B_{12}N_1/B_{21}N_2)-1]$

This by Boltzmann above gives:

$$\rho(f_{21}) = (A_{21}/B_{21}/[(B_{12}/B_{21})\exp(\Delta E/kT)-1]$$

Compare this with Plank above and we must conclude:

$$B_{12}=B_{21}=B \text{ simply and also } A_{21}/B_{21}=A/B=8\pi hf_{21}^3/c^3$$

We also see that the ratio spontaneous/stimulated is:

$$A_{21}/\rho(f)B_{21} = [\exp(hf/kT)-1] = \sim \exp(hf/kT)$$

At 10^{14} Hz in the IR this produces

$\exp(6.6x10^{-20}/4.2x10^{-21}) = \sim \exp(16) \text{ or } 10^5$.

We conclude spontaneous emission is enormously more likely than stimulated in the IR. Hence the historical difficulty of getting laser action.

Conversely at rf of 10^8 Hz we get $\exp(16/10^6) = \sim 1$ so the ratio above is $(1-1)=0$ meaning stimulated emission is certain for radio waves. A simple antenna is sufficient.

Exercise: At what frequency will the competing processes be equiprobable?

Answer: Put ratio = $A_{21}/\rho(f)B_{21}$ = 1 to get exp(hf/kT)=2 and thereby hf/kT= ~1. Hence hf/kT= ~1 or hf=kT

Now f = kT/h = $1.4 \times 10^{-23} \times 300/6.6 \times 10^{-34}$ = ~100 GHz in the far-IR or well beyond fibre optic windows.

Laser Threshold

The dh design is also used for semiconductor lasers and the prior Figure 13 is a typical case except that there is feedback from the end facets which are now cleaved to give a mirror finish and the device is driven above the threshold current I_t whose value will be determined next. The lasing cavity is formed by the junction which is along the <1,0,0> planes of the wafer. The orthogonal planes <1,1,0> are cleaved to produce equal reflectivity given by Fresnel where n=3.6 and n_{air}=1:

$$R_1 = R_2 = (n-1)^2/(n+1)^2 = 0.32$$

This 32% is more than sufficient feedback for laser action. Above I_t stimulated emission exceeds spontaneous and the LED behaves as a laser, Figure 16. The axial path then dominates whereas the LED operation had been omnidirectional. This gives a highly *directional beam* at the output except that at the narrow exit "slit" severe diffraction takes place and collimation with a lens may be required. Above I_t also, there is *spectrum narrowing* over ten-fold while uniformity of phase provides a *coherent beam* that exhibits *polarised cavity modes*.

Threshold is the point at which amplification of preferred modes by stimulated emission begins to exceed combined losses by absorption, scattering and optical losses at the output facets. In electronics this is where the oscillator is said to have closed loop gain of unity.

We will simplify the chip, Figure 19, in order to put the above statement of threshold into mathematics.

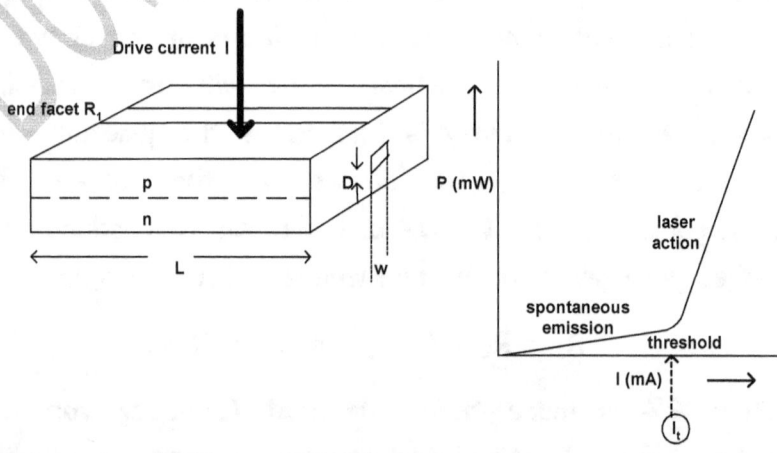

Figure 19. (a) Facets are separated by chip length L to form the lasing FP cavity. (b) Optical power versus drive current has a knee at threshold I_t.

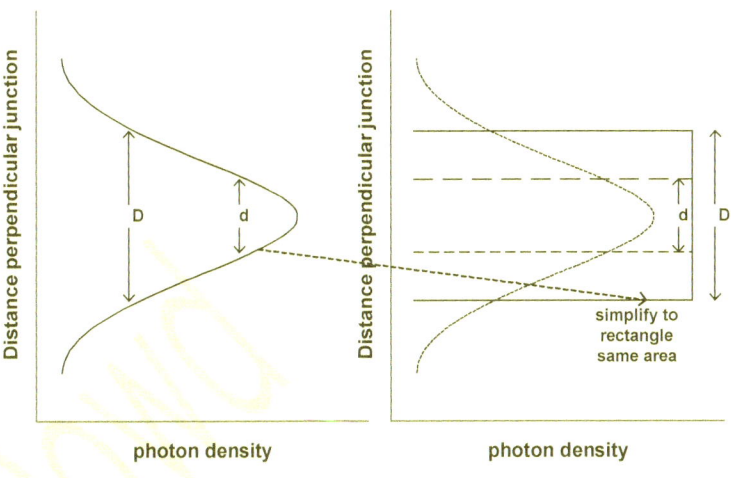

photon density   photon density

Figure 20. The near field with half power width D is simplified from bell to rectangle; d is active layer depth.

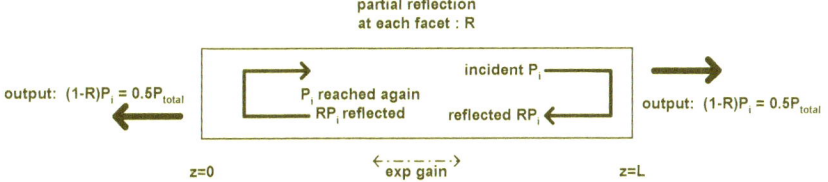

Figure 21. Light path up/down the cavity from z=0 to L. Total power is double the single facet output.

The light output at each chip end has a bell shaped near field intensity pattern that we simplify to a rectangle, Figure 20. Some light spreads out to extent D. The photons inside the active region of width d get amplified while those outside do not. All photons can suffer loss.

These exponential processes are summarised in Figure 21 to give power $P(z)$ at depth z inside the chip:

$$P(z) = RP_i\exp[(gd/D-\alpha)z]$$

Here the power P_i at a facet has portion R reflected at $z=0$ and that portion d/D is amplified at gain rate g while all light can suffer loss at rate α along the z-axis. For unity round trip gain RP_i recovers to P_i at $z=L$:

$$P_i = RP_i\exp[(g\Gamma-\alpha)L]$$

We use $\Gamma=d/D$ as confinement factor for the light restricted within the gain region. This produces:

$$\ln(1/R) = (g\Gamma-\alpha)L \text{ or } g\Gamma = \alpha + (1/L)\ln(1/R)$$

Use $g = \beta J$ where from physics $\beta = q_e\lambda_o^2/8\pi en^2 d\Delta v$

to give $\qquad\qquad \beta J_t = \alpha + (1/L)\ln(1/R)$

$I_t=Area.J_t$ hence $I_t = (wL/\beta)[\alpha+(1/L)\ln(1/R)]$

J is injection current density (Am^{-2}) that provides and is proportional to gain g with constant β being device and material dependent.

q_e is internal quantum efficiency as before

λ_0 is vacuum wavelength at centre of laser gain spectrum

Δv is spontaneous emission spectrum width

n is gain medium refractive index

Exercise: Show by repeating the I_t derivation that when the facets have unequal reflectivity we use $(R_1R_2)^{1/2}$ in place of R in the above equation for threshold.

You Do...

Ex 5. Threshold current.

Find J_t and I_t for a GaAs laser of length 200 μm, injection stripe width 10 μm, uncoated facets. refractive index 3.2, confinement factor 0.8 where α = 10 cm^{-1} and β = 2.0x10^{-4} m/A

Answer Ex 5 at end handbook.

For Γ=1: J_t=3.35 kAcm^{-2} and I_t=67 mA

For Γ=0.8: J_t=3.35/0.8=4.19 kAcm^{-2} and I_t=83.7 mA

You Do...

Ex 6. External feedback.

R for the laser is effectively increased by 10% due to optical feedback from a nearby optical fibre facet. Re-calculate I_t for the above exercise.

Answer   Ex 6 at end handbook.

○

You Do...

Ex 7. L-I or P-I plot.

Illustrate the effect on the P-I plot for this laser with and without the fibre feedback.

Answer   Ex 7 at end handbook.

○

Laser Temperature Effects

At higher temperature T the threshold rises as shown in Figure 22 and the result is we can get "clipping" of the optical intensity modulation for the same bias current I_b. A higher bias added to the signal modulation is required to prevent this. A better alternative is to stabilise temperature with a Peltier-effect cooler. This is commonplace in laser transmitters with temperature control within +/- 0.01 K. The Peltier-effect uses a thermo-couple type of device working in reverse; current across the bi-metal junction cools it.

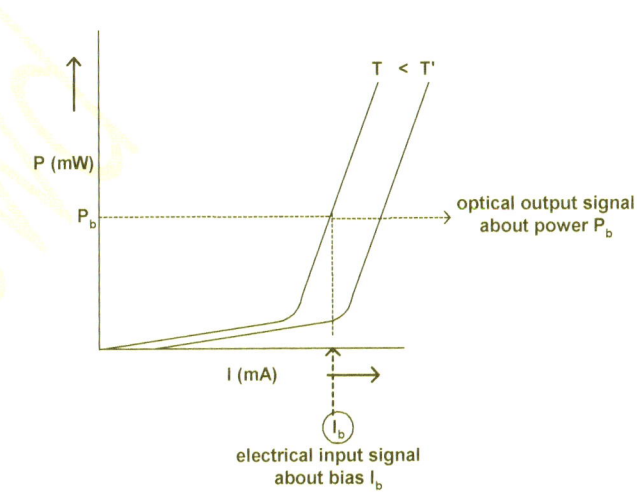

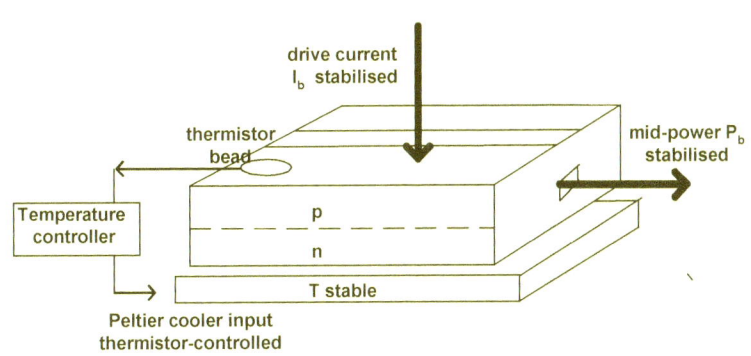

Figure 22. Upper: The P-I (or P-J) curve moves to higher current with increasing temperature T(K). Lower: Peltier-cooler stabilised laser chip.

$J_t(T)$ is influenced by variation in efficiency q_e, electrical and optical confinements and other factors. It is usual to fit an exponential to the combination as is expected of all thermal effects:

$$J_t(T) = J_0 \exp(T/T_0)$$

Here J_0 and T_0 are empirically determined coefficients.

Hence

$$\left(1/J_t\right)dJ_t/dT = 1/T_0$$

We conclude that for low variation with temperature a *high T_0 is desirable*. Typically it is 150 K for ternary dh lasers but only 70 K for quaternary material due to q_e falling off faster with rising temperature.

A further effect is the wavelength variation with temperature $d\lambda/dT$, about 0.3 nm/K so that temperature regulation using a thermistor and Peltier cooler combination is essential for stable spectrum.

Laser Rate Equations

We now consider the two-level case applicable to semiconductor lasers where the dynamics are governed by coupled rate equations, one for carrier inversion density n and one for photon density s in the gain region. (Note: this is *not* the n also used for refractive index).

$$dn/dt = I_b/eV - Bns - n/\tau$$

$$ds/dt = Bns + F(n/\tau) - s/t_p$$

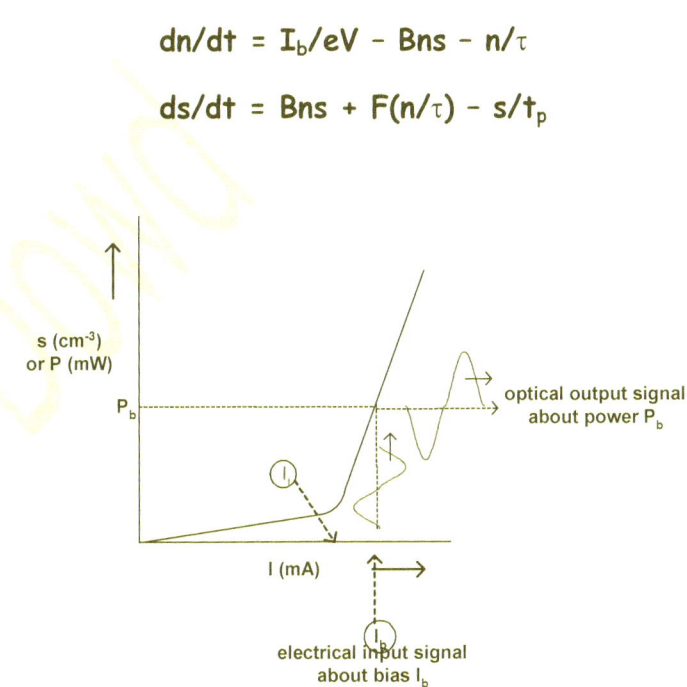

Figure 23. Plot of photon density s (cm^{-3}) versus injection current I (mA). This s mimics optical power P_0.

The first equation is the variation with time of injected carrier density due to bias current I_b in volume V where stimulated emission Bns and spontaneous recombination n/τ are meantime depleting the population. The *carrier lifetime* before spontaneous recombination is τ so the rate for this process is n/τ. The second equation gives variation with time of photon density where light is created by stimulated emission Bns and by fraction F of omnidirectional spontaneous emission F(n/τ) launched into the coherent direction. Simultaneously light s/t_p is lost at the end facets at a rate determined by t_p which is the average *photon lifetime* within the cavity before exiting. The common term that couples these differential equations is stimulated emission Bns as it is dependent on both variables n and s. The "pump" term is I_b/eV being the injection into volume V of carriers each of charge e. The fraction F is small, order 1%, so we can neglect it to see what is happening to n(t) and s(t).

Under the continuous wave output or "cw condition" ds/dt is zero giving:

$$n_t = 1/Bt_p \qquad \text{threshold carrier density}$$

Under the "threshold condition" and above it the carriers are at equilibrium density n_t so dn/dt is zero giving from the second differential equation along with known n_t:

$$s_b = I_b t_p/eV - 1/B\tau = (W - n_t/\tau)t_p$$

Here s_b represents coherent light output at the bias point I_b in the plot. W or I_b/eV is injection carrier density rate at the selected bias while n_t/τ is spontaneous recombination. The last equation subtracts spontaneous "waste" from the total and converts to coherent light output.

Example: Work out how to convert from s the photon density in photons per cm^3 to optical output power P_0 in watts. Hint: use Plank's equation and think of the "lit" inside volume V moving out into air at light group velocity c_g. We may conclude P_0 and s are therefore interchangeable.

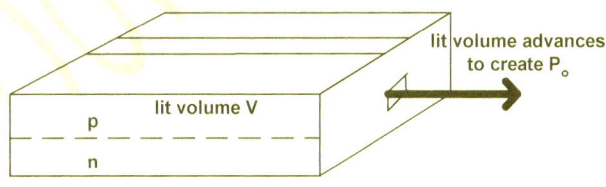

Answer: P_0 (watt) = s(hc/λ)Vc_g

Observe that for the plot of the last equation, s versus I_b, shown in Figure 23 and 24, we expect slope m in y = mx + C to be t_p/eV. Therefore we can *evaluate photon lifetime* t_p by measuring the slope of an experimental power versus current graph. It is typically some picoseconds while τ is a few nanoseconds. The intercept C is 1/Bτ so that measurement provides τ. For these P-I experiments (sometimes called L-I with L for light) we require to know device active volume V=wDL in Figure 19 and material coefficient B from Table 3.

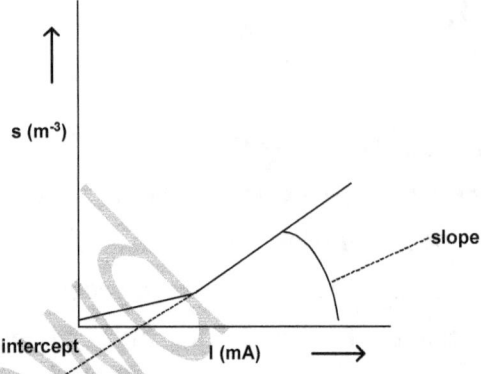

Figure 24. Slope and intercept for the s-I or P_0-I or L-I graph give photon and carrier lifetimes t_p and τ.

The lasing process can now be interpreted. At and above I_t or n_t a dynamic equilibrium is established that fixes n at value n_t so that dn/dt is zero (threshold condition). Additional pump then produces greater output, power rising with drive current, while carrier inversion remains fixed.

We may rewrite the s_b equation for a higher bias I_b'. Then consider modulating ($I_{mod} = I_b - I_b'$) between I_b and I_b'. Now subtract these two equations to get modulated light s_{mod}:

$$s_{mod} = s_b - s_b' = I_{mod} t_p / eV$$

Hence the optical *intensity modulation index* or relative light intensity variation s_{mod}/s_b is:

$$m = I_{mod} \; t_p/eVs_b$$

Clearly a longer photon lifetime t_p or smaller active chip volume V improves data modulation capability.

You Do...

Ex 8. Photon density from power.

A typical communications laser emits 10 mW or 10dBm but a high power device is selected for 20 dBm as the link is atmospheric not fibre optic. It lases at 1.5 μm and the semiconductor alloy has average index 3.45 across the lasing spectrum. The device dimensions are stripe w=10 μm, lit depth D=1 μm, chip length L=100 μm. Calculate the photon density s.

Answer Ex 8 at end handbook.

○

4 Advanced Lasers

We have seen that temperature variations can affect the spectrum of the laser transmitter and that optical *frequency stability* is achieved using a Peltier cooler. More advanced lasers involve complex design that adds embedded optical control, for example using an in-chip diffraction grating. This operates somewhat like a conventional grating with many lines, not on glass however but etched into the semiconductor surface as corrugations. The "line" spacing in the grating section determines what wavelength is reflected back into the active section of the laser chip, Bragg's Law. Furthermore, when a current is supplied to the new section the charge density rises so the effective refractive index changes and with it the wavelength. This is then an electronically *tunable laser*. These complex features comprise advanced lasers for communications that can be deployed to exploit the full bandwidth of optical fibres shown in Figure 25. This shows the conventional C-band around 1550 nm wavelength centred on optical frequency 200 THz. There is also a dip near 1300 nm or 1.3 μm that is used for computer local area networks, LANs.

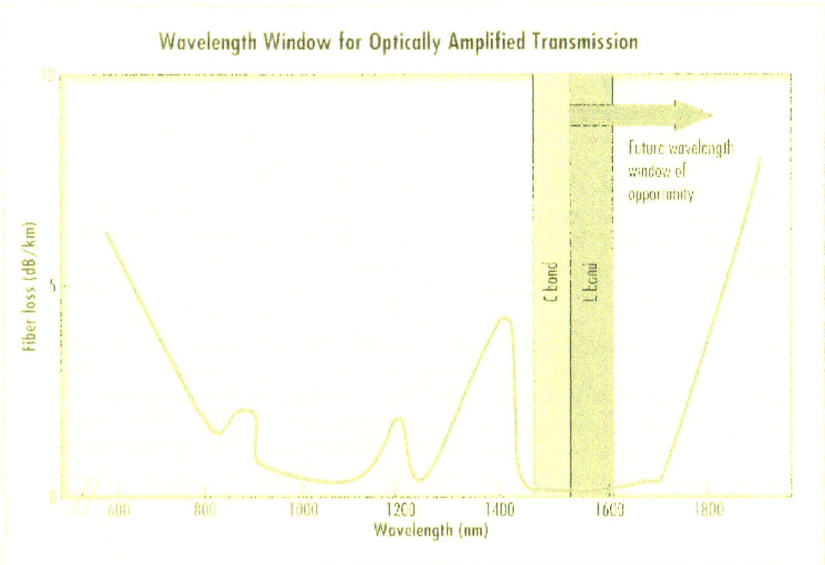

Figure 25. Transmission windows for SiO_2 or silicate optical fibre reside in the low attenuation parts of the loss spectrum. The conventional C-band at 1550 nm accommodates 50 or 100 dense optical frequency channels for DWDM.

The silicate glass made of extremely pure SiO_2 has tiny residual water contaminant that absorbs via the OH bond at 1400 nm in the near IR. This peak separates the short or S-band from the low loss *conventional or C-band* and beyond that lies the long or L-band. The C-band around 1550 nm wavelength is broad enough to accommodate 50 or 100 closely spaced optical channels. This practice is called *dense wavelength division multiplexing* DWDM. A frequency stable and tunable laser allows us to access all of this bandwidth. The optical frequency here is $\nu=c/\lambda$ or $3\times10^8/1.5\times10^{-6}$ giving 200 THz. Compare with microwave frequencies, say 10 GHz, and the bandwidth is 20,000 times higher. Then multiply by 50 channels gives 10^6 times greater capacity.

Clearly grating based lasers are needed to exploit this information usage capability and we discuss that technology next. The same devices are applicable to *coherent optical sensing* where the frequency or phase changes of the light are monitored to detect tiny variations in a property of matter.

DFB Laser

This incorporates a Bragg grating embedded as corrugations in the layer above the active region of the laser chip and it replaces the facet reflector for feedback, Figure 26. Light leaks across the thin 0.1 μm layer to the grating where constructive interference by reflection occurs subject to the Bragg Law, integer times wavelength equals path difference. As the grating is spread along the length of the chip we have a DFB or *distributed feedback Bragg* construction.

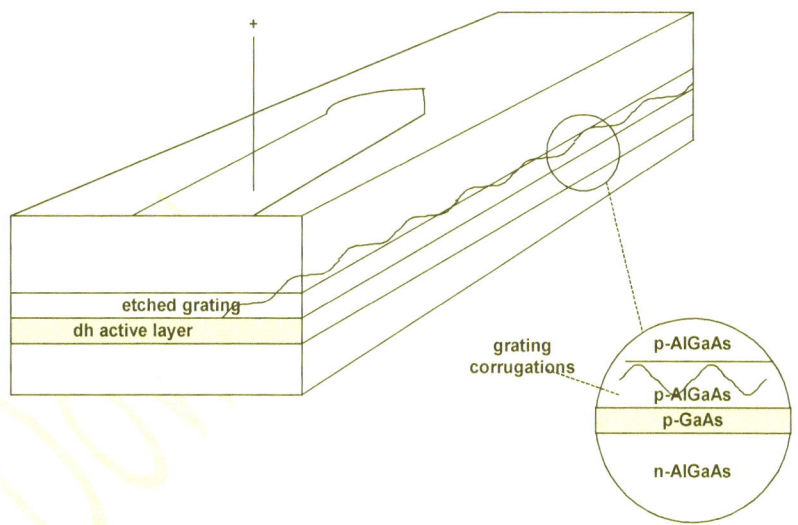

Figure 26. DFB laser structure.

The Bragg condition for reflection is:

$$N\lambda = 2n_e D$$

Here N is an integer, n_e is effective refractive index in the medium and D is corrugation period (separation of "lines"). In first order N=1, typical corrugation spacing is 0.25 μm and as index is about 3 we get λ=1.5 μm. The spectrum is now dependent on feedback according to this equation and frequency stability is improved ten-fold. The light FWHM spectrum spread $\Delta\lambda$ is <1 nm and varies only 0.05 nm/K. Compare a Fabry Perot or FP device with two-facet reflection having $\Delta\lambda$ > 2 nm and temperature dependence 0.5 nm/K.

DBR Laser

An alternative design can have gratings at the end regions only, Figure 27. This is the *Distributed Bragg Reflector* variation.

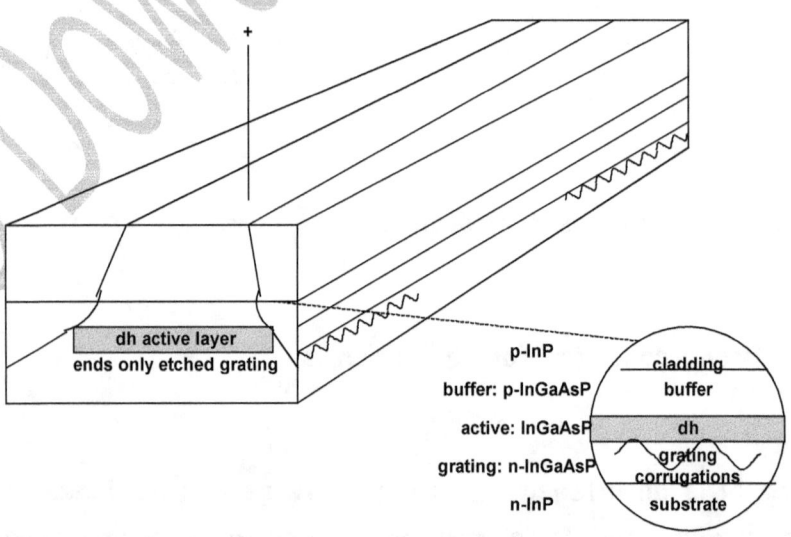

Figure 27. DBR laser.

High Power Devices

Normal lasing *spot-size* or near-field is typically 3 μm wide by 0.6 μm high while active layer thickness may be only 0.15 μm so that >50% of the light travels in the outer cladding. Such devices are limited to order 5 mW continuous-wave or cw power corresponding to 10 mW peak when intensity modulated with data (50% on). Higher power would damage the semiconductor end facet so it no longer acts as a mirror.

The TAL and LOC structures, Figure 28, are devised to overcome this limitation.

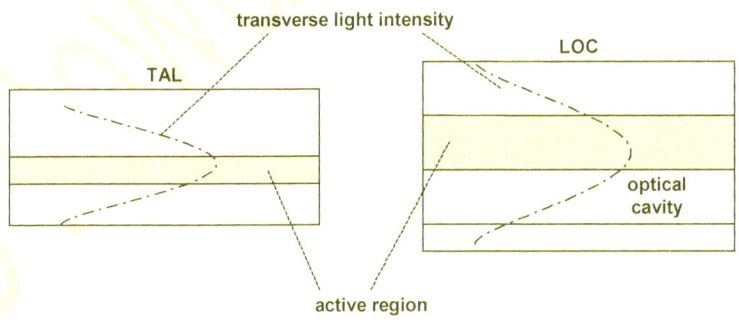

Figure 28. TAL and LOC structures for high power.

The thin active layer, only 0.05 μm, of a TAL laser causes the wavefront to diffract outwards thereby doubling the transverse spot to >1μm so 90% of light travels in the cladding. This provides up to 25 mW cw.

The large optical cavity or LOC has a large guide of intermediate index adjacent to the active layer providing a spot ~1.5 μm with power ~40 mW cw. The large spot also reduces the far field beam which is determined by diffraction. A conventional AlGaAs spot of 3x0.6 μm gives an elliptical beam 12 deg × 45 deg. The large spot device produces ~7 deg × 25 deg.

Further techniques include (1) anti-reflection AR mirrors using multi-layer dielectric coatings to the facets and (2) non-absorbing mirrors of wide band-gap that can increase power x5 times again to hundreds of mW.

Phased Array Laser, PAL.

It is possible to grow multiple lasers side by side on the same chip so that the beams combine to a powerful single output, a *phased array laser* Figure 29. The stripes at the output facet can couple via evanescent waves, analogous to a classical transverse diffraction grating where the phases lock together.

Light passing through each single slit produces a diffraction pattern, spread $2\lambda/s$, which is wider for a narrower stripe s. The multiple beams then interfere according to their separation period D with spread of the pattern λ/D. The product of these two patterns provides the output laser beam. As s is made closer in value to D the diffraction envelope encompasses only one or two interference lobes so we approach a single spot. For example a PAL with N=10, λ=850 nm and D=10 μm can have a FWHM for the central lobe about 0.5 deg and two lobes separated by ~5 deg. An array of 40 emitters with I_t=30 mA each or 1.2 A total can produce 2.5 W from a crystal only 0.15 mm³ and internal lit volume of just 10^{-5} mm³.

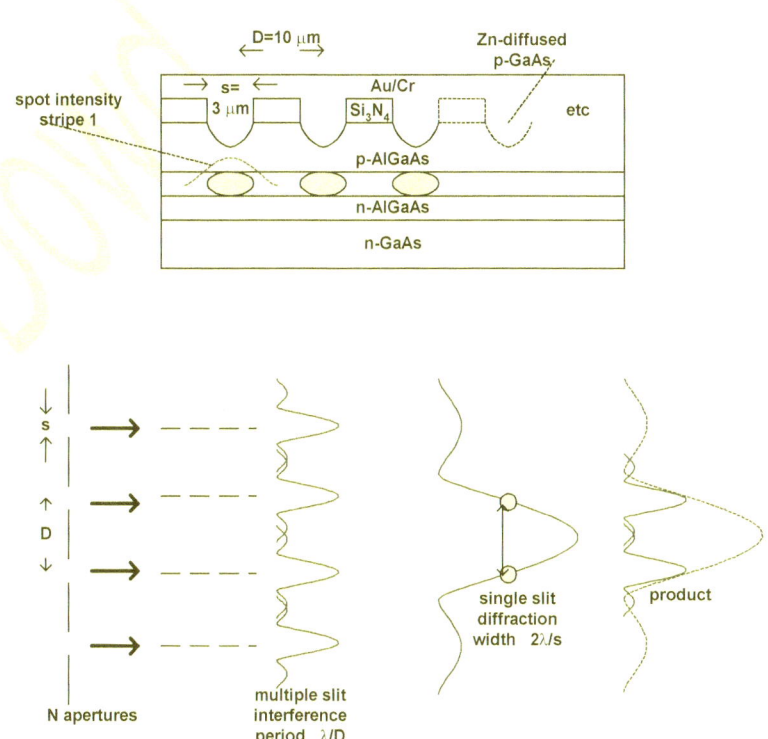

Figure 29. Multiple stripe phased array laser. Each stripe behaves like a slit and light from the N apertures diffract and then interfere.

Exercise: Calculate internal photon density s using the data for the last described laser.

Quantum Well Lasers

The potential well for an electron in the hydrogen atom is known from quantum mechanics and a similar arrangement, a *quantum well laser,* can be created artificially using a very thin layer for the active region within a semiconductor double heterostructure Figure 30.

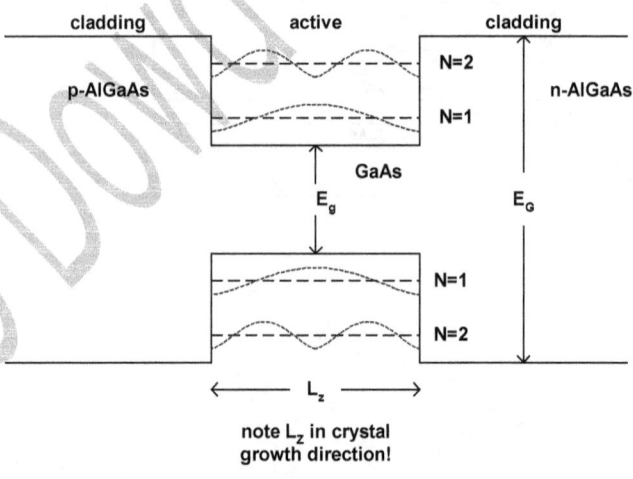

Figure 30. Electrical potential in a quantum well laser.

The width Lz of the thin active layer is chosen to be compatible with the electron wavelength λ_e as given by De Broghlie. Observe that the layers are drawn sideways in Figure 30. A conventional dh has gain region thickness ~0.2 μm but for the quantum nature to dominate it needs to be twenty time smaller, ~10 nm or 100 angstroms. Then the electron cloud is effectively a two dimensional gas along the hetero-boundary so there is very high electron mobility. (For that reason the same structure has migrated from photonics into the design of a high electron mobility transistor or HEMT for use in microwave communications). For the electron standing wave in the thin active layer as illustrated, an integer times half wavelength must equal the physical space so:

$$N \lambda_e /2 = L_z$$

Use from physics electron momentum $p=h/\lambda_e$ and also motional energy $E_Q = p^2/2m^*$ where m^* is the effective electron mass.

Exercise: Show that these relations along with the equation above will produce separation ΔE_Q of the quantised energy levels. Hint: use N=1 and then N=2 and subtract.

Answer: Combine the three relations to get:

$$E_Q = N^2 h^2/8m^* L_z^2$$

Insert N=2 and N=1 and subtract for sub-level spacing:

$$\Delta E_Q = 3h^2/8m^* L_z^2$$

This gap relates to the closely-spaced stack of energy levels shown in Figure 31 for bulk and then QW material.

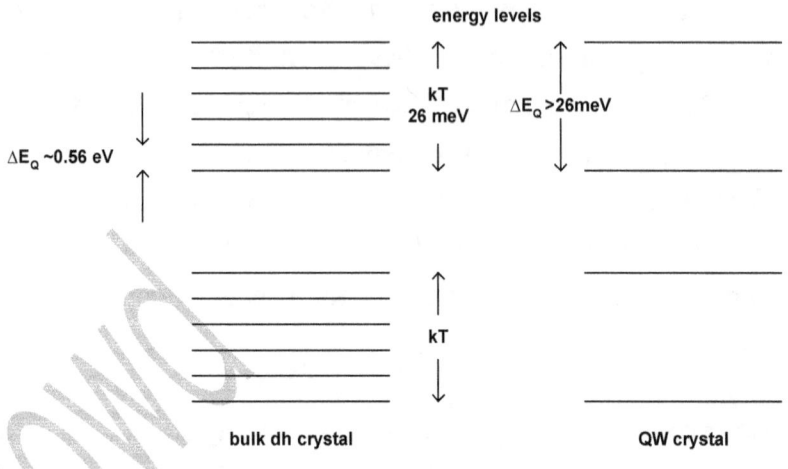

Figure 31. Energy levels in bulk dh material, left, and artificial QW crystal, right.

When known constants and L_Z <20 nm are inserted into this result we find ΔE_Q which was 0.56 meV in a conventional or bulk dh laser now becomes >26 meV for the quantum well device, Figure 31. This result is crucial as it means the levels are now separated by more than the thermal energy kT=26 meV at room temperature. That value of kT was large enough to encompass the spacing of many close-packed levels in the *bulk material* but in artificial QW medium the levels have spread to the new ΔE_Q so only about two are occupied at room temperature. Population inversion is now achieved with ease and the threshold current I_t falls by decades. That in turn means less heating in the tiny chip, less spectral deterioration and reduced temperature dependence.

The smaller dimensions mean we require *nanotechnology* to grow the QW lasers and also diminished active volume suggests using several quantum well layers in a stack to get cumulatively enhanced optical power. This is a *multiple quantum well or MQW* laser. Outside the stack we grow dh layers to contain the light and carriers, the so called *separate confinement heterostructure SCH*. The nanotechnology may be achieved with advanced crystal growth such as metal-organic vapour phase epitaxy, MOVPE.

Vertical Cavity Surface Emitting Lasers, VCSELS

Although edge emitters are widespread in fibre optic communications there has been a return to surface emitting technology for CD players, laser printers etc, since the advent of QW capability. These VCSELs (pronounced "vicsels"), illustrated in Figure 32, have smaller spot size and narrow beam so they also can be more readily focussed into a fibre.

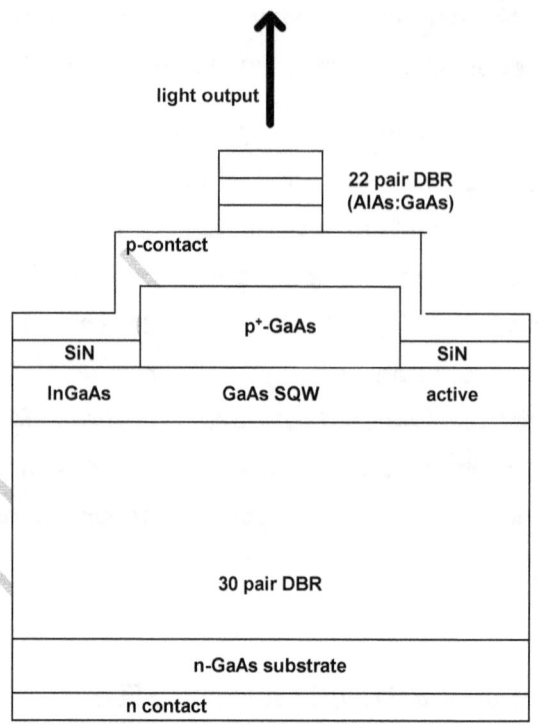

Figure 32. VCSEL with DBR mirrors.

The mirrors of the FP cavity are above and below the single quantum well, SQW, light-creating region. The mirrors can be made of multiple semiconductor layer-pairs of high and low index to form DBR reflectors. These pairs replace the corrugations of prior bulk DBRs. This VCSEL device is a triumph of nanotechnology and experiments have also shown it can be deployed at above 100 Gbit/s data rate when the spin of injected electrons is polarised. A great advantage of surface emitters is the individual lasers may be inspected for L-I plot before dicing into chips; wafer inspection is more economical than chip-by-chip for mass production.

Nanotechnology, Artificial Silicon, Optical Computers

The quantum well already described can be extended to two dimensions creating a quantum wire, or even to 3D resulting in a quantum box within the crystal called a *quantum dot*. The threshold current is thereby reduced significantly with even lower power consumption. But there are further benefits to these artificial crystalline materials. An electron may orbit a hole similar to a hydrogen atom with a nuclear proton. This carrier pair in a semiconductor is an *exciton* and in such a state the electron is localised to a position x with less room Δx to manoeuvre. By the Heisenburg Uncertainty Principle $\Delta x \Delta p$ relates to the Plank constant h so smaller Δx means greater Δp. That implies, when considered in the context of Figure 4 for indirect band gap silicon, that a significant momentum shift Δp is more probable facilitating a luminescent recombination even in Si. Now consider a Si crystal grown with a matrix of quantum dots. As electrons will be trapped in the quantum well boxes they are localised so the artificial crystal facilitates radiative recombination. The dream of using the most common element in the Earth's crust, Si, for lasers can become a reality.

These materials also exhibit a refractive index n that varies with light intensity I so that higher order $n(I)$ terms must be added to our equations and a vista opens for *non-linear optics*. This provides for parallel photonic processing using optical wavefronts and for ultrafast optical computers.

Tunable Laser

The DBR device depicted in Figure 26 has an oscillating wavelength governed by Bragg's Law:

$$N\lambda = 2n_eD = f(n_e, D) \qquad \text{Bragg's Law}$$

This indicates we can control the light by both etching the corrugations in the end grating reflector to the correct period D and by effective refractive index n_e. This latter is determined by the alloy in the reflector layer but there is a further dependency; an injected current can raise the carrier density, changing n_e, and that will cause the feedback wavelength to shorten. This means we have an *electronically tunable laser* with a gain current I_A to the active layer section for power setting and a tuning current I_G to the DBR grating section for frequency control. A third section for optical-phase control can also help, and indeed a fourth section as

DBR reflector at the other chip end. This construction was devised by the team led by Coldren at the University of California; these controls are the phase current I_P and second grating I_{G2}. The combination $(I_A, I_{G1}, I_{G2}, I_P)$ was found to give tunability across a full quarter of the C-band so that only four lasers in the transmitter TX could cover the entire ITU C-band comb of 100 optical channels. This opens the way to agile optical frequency allocation and consequently *"bandwidth on demand"* for customers of Internet Service Providers. The detailed operation of this laser is left as a Library Exercise (see O'Dowd, IEEE Jnl Selected Topics in Quantum Electronics, Special Issue March 2001).

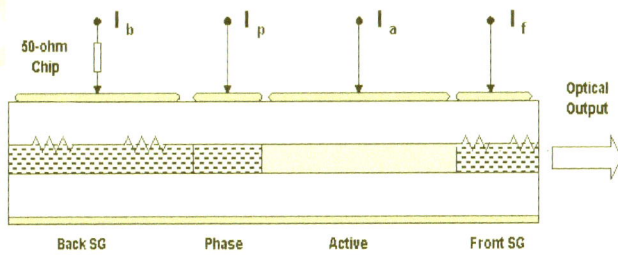

Example of a tunable laser with back and front gratings and four tuning currents. (Coldren et al, UnCalSB).

This type of tunability also permits passive *wavelength routing* of data through a network since the tuned IR-colour can be the packet's address and optical filters, like gratings, at each node can direct it do destination.

5   Photodetectors

Silicon in its common or natural form was forsaken in our search for luminescence but it is certainly useful in receivers for the detection of light as long as the wavelength is shorter than 1100 nm corresponding to the energy gap. Si is transparent to and absorbent of higher frequencies. This restricts it to UV, visible and very near IR. At longer wavelengths, especially C-band, we require III-V compounds in photodiodes.

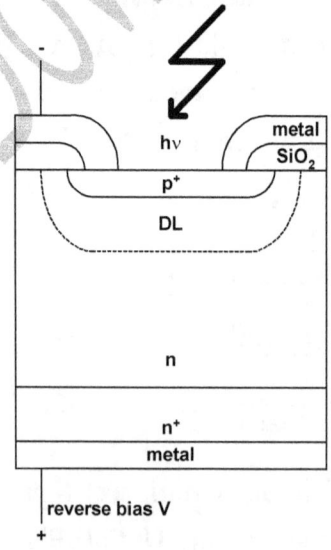

Figure 33. Silicon pn diode photodetector.

The simplest structure is a p^+n diode having a heavily doped p^+ side, Figure 33. With a p^+ layer in top of the n region the depletion layer (DL) extends well into the n side as shown but is compacted on the p side. Below there is an n^+ substrate for good ohmic contact with the metal. A window above in the SiO_2 silicate insulator allows light photons, $h\nu$, to enter. After photon absorption and under *reverse bias* the electron hole pairs are created within a diffusion length of DL and swept out to provide a photocurrent. The absorption coefficient is high at shorter wavelengths. At longer wavelength λ a wider DL is needed to ensure absorption implying higher reverse bias V but eventually that will exceed the breakdown voltage. The solution is a p-i-n structure or PIN diode with a wide intrinsic i-layer, Figure 34.

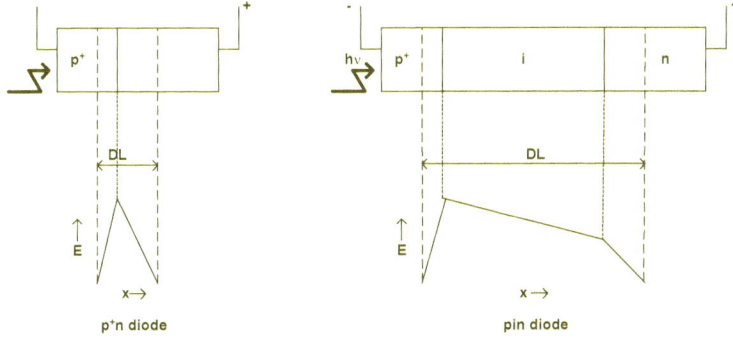

Figure 34. PIN diode with intrinsic Si central region. The internal E-field extends through the i-layer.

The idea here is that intrinsic silicon being un-doped has relatively high resistivity so the internal E-field falls slowly as depicted and there is effectively a wide absorption volume created as photons penetrate this. Only a few volts are required for DL to penetrate right through so that common logic processor levels are sufficient, e.g. 5 V for TTL. This design is useful at the longer wavelengths where the absorption coefficient is small. An additional benefit is that the capacitance associated with the junction, $C=\varepsilon A/d$, is diminished at longer d. That provides for shorter lifetime, $\tau=RC$, *faster response* and therefore wider bandwidth. The operation whereby a photon creates an electron-hole pair that is swept out by the internal E-field is depicted in the energy level diagram, Figure 35.

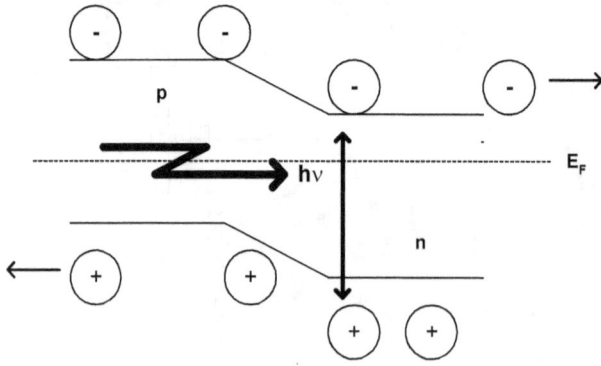

Figure 35. Energy levels in a photodiode and photon absorption creating electron-hole pairs that are swept out by the internal E-field.

On open circuit a potential difference appears at the output terminals. This is called *photovoltaic mode PV*.

When an external diode reverse bias V_d is used a photocurrent is available internally i_λ and portion i_{ext} is available at the terminals, Figure 36. This is called *photoconductive mode PC* and is most common in communication receivers, RX.

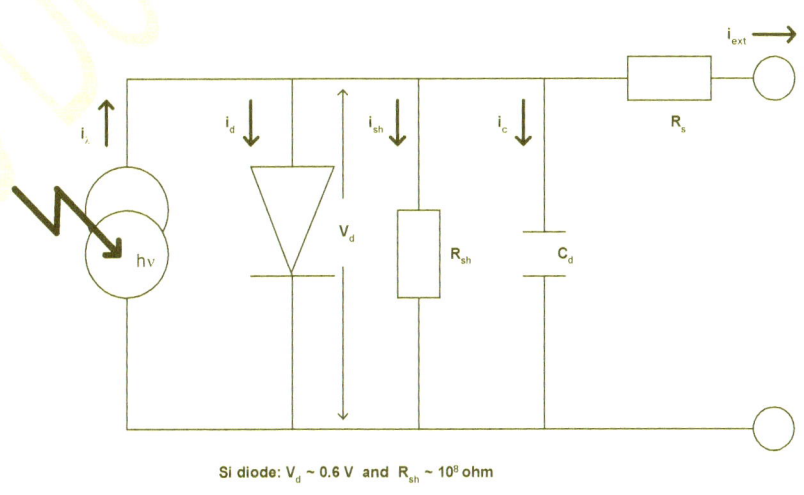

Si diode: $V_d \sim 0.6$ V and $R_{sh} \sim 10^8$ ohm

Figure 36. Equivalent circuit for photodetector.

If all incident radiation is absorbed within the photocell then each photon $h\nu$ creates an electron in the photocurrent i_λ:

$$i_\lambda = q_d(I_{opt}A)/h\nu)e$$

Here q_d is the diode efficiency, A is the window area, e electronic charge and I_{opt} is the optical energy density arrival rate called *irradiance* with units $Jm^{-2}s^{-1}$ or Wm^{-2}. For example sunlight irradiance at European latitudes is $\sim 1\ kWm^{-2}$.

Sensitivity

The equivalent circuit in Figure 36 has the photocurrent generator at left delivering current i_λ that breaks up into diode current i_d, shunt resistor current i_{sh} and diode capacitance current i_c. Only remainder i_{ext} flows through the series resistance R_s and is available externally for amplification and processing. In first evaluation of the photodiode *sensitivity* neglect the model's current i_C through the capacitance, valid at low modulation frequencies. Then apply Ohm's law, Kirchoff's law and the common diode equation as follows:

$$i_\lambda = i_d + i_{sh} + i_{ext}$$

$$V_{ext} = V_d - R_s i_{ext}$$

$$i_d = i_0[[(exp(eV_d/kT)-1]$$

$$V_d = R_{sh} i_{sh}$$

In PV mode i_{ext} is very small and can be neglected by comparison giving:

$$i_\lambda = i_d + i_{sh}\ and\ V_{ext} = V_d$$

Hence $\qquad i_\lambda = i_0[\exp(eV_d/kT)-1]+V_d/R_{sh}$

$$\exp(eV_d/kT) = 1+i_\lambda/i_0-V_d/R_{sh}i_0$$

For a silicon diode V_d=0.6 V and R_{sh}=10^8 ohm . Typically reverse leakage current in the diode equation is only ~10 nA so the last equation simplifies to:

$$\exp(eV_d/kT) = i_\lambda/i_0 \text{ since } i_\lambda>>i_0$$

Natural log of each side produces:

$$V_d = (kT/e)\ln(i_\lambda/i_0)$$

Finally using V_{ext}=V_d for parallel branches along with the original photocurrent i_λ equation gives:

$$V_{ext} = (kT/e)\ln(q_d I_{opt}Ae\lambda/hci_0) \quad \text{PV MODE}$$

This tells us the external voltage in PV mode is a log function of incident light irradiance. That mode is used in solar cells and the power resource for that at 50 deg latitude is $I_{opt} \sim 1$ kW/m^2.

PC mode

Typically ~10 V reverse bias is applied to operate in photoconductive mode for telecommunications links so the diode is well into the reverse saturated characteristic and i_d=i_0 giving from the first equation for i_λ:

$$i_\lambda = (i_0+i_{sh})+i_{ext}$$

The bracketed current is termed the "dark" current as it is present even when no light or data signal is present. It represents noise in the receiver. Again for a Si diode $i_{sh}=V_d/R_{sh}=10/10^8=100$ nA while i_0 is ~10 nA and i_λ is typically ~1 μA. This order of magnitude comparison of terms allows us to simplify:

$$i_{ext} = i_\lambda = q_d I_{opt} Ae\lambda/hc \qquad \text{PC MODE}$$

In photoconductive mode the external current available for processing is directly proportional to the incident light irradiance I_{opt}. This means an intensity modulated IM laser will provide a linear response at the RX. We see that response is also proportional to wavelength λ.

The PC mode has a *linear response* to light signal and is also faster, has better stability and greater dynamic range than PV mode. The *dark current* however (i_0+i_{sh}) gives rise to *shot noise* which limits ultimate receiver sensitivity.

Responsivity of Si

The *responsivity* is depicted for silicon in Figure 37 showing no response to light beyond 1.1 μm as the band gap there is too large to absorb these smaller photons.

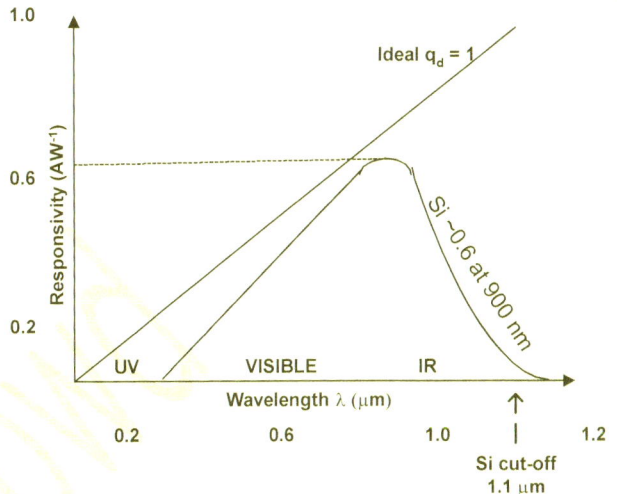

Figure 37. Responsivity spectrum for Si photodiode. Other materials have similar shape, different cut-off.

Photocurrent i_λ was defined using diode material efficiency q_d and optical irradiance I_{opt} in Wm^{-2} as follows:

$$i_\lambda = q_d(I_{opt}A/h\nu)$$

We define photodiode *responsivity* as $i_\lambda/I_{opt}A$ or photocurrent in amp produced per watt of received light.

$$\text{Responsivity} = i_\lambda/I_{opt}A = (q_d e/hc)\lambda \quad \text{units: } (AW^{-1})$$

The ideal efficiency is unity (or 100%) but in practice it can be ~0.6 at 900 nm when using Si. That means the ideal straight line plot shown in Figure 37 is not attained using Si but rather is about 60% lower stretching from the UV at 300 nm to 1100 nm in the near IR where the band gap cuts it off rather abruptly. The units of responsivity are often quoted in $mAmW^{-1}$ to reflect actual currents achieved in a fibre optic link and crucially we see it is proportional to wavelength... *the responsivity grows with* λ. The cut-off is at 1.1 μm for Si but 1.8 μm using germanium Ge and lies beyond 1.6 μm for selected III-V alloys so they can be deployed for C-band systems where their responsivity is closer to ideal.

Exercise: Responsivity dependence on wavelength.

Try to explain why the response should rise linearly with wavelength in the ideal photodiode case.

Hint: estimate number of photoelectrons created for each watt of received light starting from Plank's law.

The properties of common photodetector materials for different photonic applications are shown in Table 5. The last column is ratio of the ionisation coefficients $k=\alpha/\beta$ for electrons to holes and is useful for the avalanche effect that we consider next.

Table 5. Properties of photodetector materials.

Material	Bandgap (eV)	Cut-off λ	Ratio α/β=k
Si	1.1	1.1 μm	10-100
Ge	0.67	1.8	0.5
GaAs	1.43	0.85	1-0.01
$GaIn_{86}As_{14}$	1.15	1.1	0.25
$GaIn_{47}As_{53}$	0.75	1.6 note!	5
InAs	0.33	3.8	This column
InGaAsP	1.34-0.78	0.92-1.6 !	electron/hole

Avalanche Photodiodes APDs

The avalanche photodiode provides internal gain of the received photocurrent and uses for this a structure similar to the PIN. High reverse bias produces avalanche multiplication in the I-region. Average multiplication factor <M> is ~10 to 100.

Automatic Gain Control AGC

Since the ionisation coefficient for electrons and holes, α and β in Table 5, are critically temperature dependent the reverse bias requires control for thermal drift. This will give more uniform avalanche gain. Also α and β are exponentially related to E-field so <M> is sensitive to voltage and that must be automatically controlled within +/- 10 mV. The gain process is statistical with a spread of multiplications for pairs of electron-hole. The ratio k of α to β in Table 5 or its inverse 1/k is selected depending on which particle is dominant in the avalanche process for that material. The variation of gain values contributes noise to the photocurrent. This results in an excess noise factor f(M) empirically modelled as follows:

$$f(M) = M[1-(1-k)(M-1)^2/M^2]$$

We see that f(M) reduces with k and is worst when k=1. In the case of Si we use 1/k in the model as holes dominate the avalanche in that semiconductor. An alternative power-law fit to experimental behaviour is often used:

$$f(M) = <M>^x \text{ where } x=0.3 \text{ to } 0.5 \text{ for Si}$$

In practice we deploy an APD when the detector *shot noise* is well below the *circuit thermal noise*. Then optimum *signal-to-noise ratio SNR* occurs when the multiplication process, controlled by the voltage, brings shot noise up to the same level as thermal noise. Further increase in M will then deteriorate SNR. This will be shown next.

Receiver SNR

Consider a transmitter TX with optical power modulated sinusoidally at frequency ω about average power P_t with a modulation index m:

$$P(t) = P_t(1+m\sin\omega t)$$

Any more complex data signal, such as a square wave, can be broken into a Fourier sum of sine waves so this analysis is equally valid. The RX avalanche photocurrent, neglecting distortion due to fibre dispersion, will be:

$$I = I_0\langle M\rangle(1+\sin\omega t)$$

Here $I_0=q_e e P_r/h\nu$ is the average signal current before multiplication (*not* irradiance I_{opt}) and P_r is the received optical power after fibre loss. With a PIN diode use $\langle M\rangle=1$ and $f(\langle M\rangle)=1$ in what follows. Neglecting multiplied dark current the total average noise in the system from electronics coursework is:

$$\langle i^2\rangle = \langle i^2\rangle_c + 2eI_0\langle M\rangle^2 f(\langle M\rangle)B$$

Here $\langle i^2\rangle_c$ is mean-squared non-multiplied circuit noise, the added term is shot noise associated with I_0, and B is effective noise bandwidth. The shot noise has $\langle M\rangle^2$ since I_0 is multiplied first by $\langle M\rangle$ and afterwards the noise on this is itself multiplied by $\langle M\rangle f(\langle M\rangle)$. Defining SNR in terms of the ratio of mean squared signal to mean squared noise currents:

$$SNR_{APD} = \tfrac{1}{2}.m^2<M>^2I_0^2/[<i^2>_c+2eI_0<M>^2f(<M>)B]$$

$$SNR_{PIN} = \tfrac{1}{2}.m^2I_0^2/[<i^2>_c+2eI_0B]$$

This is because averaging the mean-squared sinwave produces $\tfrac{1}{2}$ at the numerator.

The amplifier circuit thermal noise is known from electronics coursework to be:

$$<i^2>_c = 4kTB.F/R_L$$

Here F is the noise factor of the input electronic pre-amplifier or low-noise amplifier LNA and R_L is amplifier input resistance.

Digital and Analogue Photodetection

For a given m, B and LNA, i.e. fixed $<i^2>_c$, the SNR equation above is a function of I_0. When I_0 is small so is its shot noise so the circuit noise dominates:

$$SNR_{PIN} = \tfrac{1}{2}.m^2I_0^2/<i^2>_c \qquad \text{Circuit noise limit}$$

This implies low SNR and therefore a *digital system* where only two binary levels need to be discerned. Conversely when I_0 is large the shot noise dominates:

$$SNR_{PIN} = m^2I_0/4eB \qquad \text{Shot noise limit}$$

This applies to an *analogue system* since large I_0 is required to discern multiple levels and faithfully reconstruct the signal shape.

As the latter expression is independent of circuit noise it applies to a perfect electronic construct (ideal LNA) so it represents the *FUNDAMENTAL QUANTUM LIMIT* for sensitivity of analogue systems or even a digital system with perfect LNA.

Optimum SNR for APD

Consider the APD expression above for SNR with initially $<M>=1$ where is behaving as a PIN. Now raise the reverse bias voltage so that SNR increases with $<M>^2$ for a while as we leave the circuit noise limit. Eventually however the growing shot noise equals circuit noise. Thereafter SNR falls off again with $[f(<M>)]^{-1}$.

Exercise: Alternatively, remembering $<M>$ depends on voltage, you may differentiate the SNR_{APD} expression with respect to $<M>$ and optimise by setting $d/d<M>$ at zero. The two noise expressions will then be equal. Use U/V derivative expression $(VdU-UdV)/V^2$

An alternative to the APD is a monolithic chip that matches a LNA to a PIN diode and because of miniaturisation exhibits overall excellent SNR. The in-built amplifier is often a field-effect transistor and this is a PIN-FET photo-receiver, Figure 38.

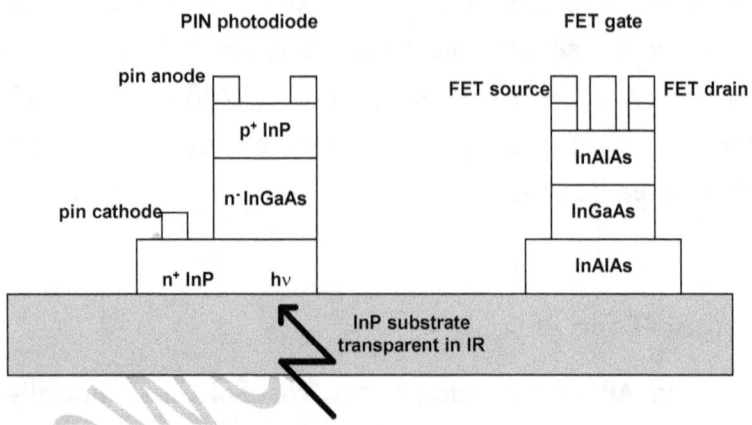

Figure 38. PIN-FET photodetector receiver.

Summary

We have learned to select photonic materials that match our requirements by band-gap engineering. Injection luminescence was deployed in various LED structures and then in semiconductor lasers. Advanced lasers were studied that cover the entire ITU C-band. Nanotechnology was perfected to deliver artificial crystals and take advantage of quantum well QW behaviour. Photodetection was modelled to get optimum SNR and linear response for fast receiver designs.

6   Answers to Exercises

Ex 1

Groups III, IV and V have the photonic materials:

III   Al, Ga, In,

IV   Si, Ge,

V   P, As

Ex 2

Calculate the wavelength associated with a 1 eV energy gap and also for Si and GaAs where it is 1.12 eV and 1.44 eV respectively.

$E_g = hc/\lambda$ gives $\lambda = hc/E_g$

At 1 eV $E_g = 1.6\times10^{-19}$ J

Hence $\lambda = 6.6\times10^{-34}\times3\times10^8/1.6\times10^{-19} = 1.24$ μm

At 1.12 eV for Si... $\lambda = 1.24/1.12 = 1.11$ μm

At 1.44 eV for GaAs... $\lambda = 1.24/1.44 = 0.86$ μm

Ex 3

A forward current of 30 mA is injected into a GaAs LED with quantum efficiency 95%. What output light power results? Use data from Ex 2.

GaAs quantum efficiency 0.95

$I = 30$ mA $= 0.03/1.6 \times 10^{-19}$ electrons per second

Optical power $P_o = 0.95(0.03/1.6 \times 10^{-19})1.44 \times 1.6 \times 10^{-19}$

$$= 0.95 \times 0.03 \times 1.44 = 0.04xx \text{ W} = 40 \text{ mW}$$

Ex 4

External quantum efficiency:

$$F = \tfrac{1}{4}(n_2/n_1)^2[1-(n_1-n_2)^2/(n_1+n_2)^2]$$

For GaAs $n_1=3.6$ while in air $n_2=1$ gives:

$$F = \tfrac{1}{4}(1/3.6)^2[1-(0.36)^2/(2.36)^2] = 1.9\%$$

Ex 5

Find J_t and I_t for a GaAs laser.

L=200x10^{-6} m   w=10x10-6 m   R=0.32 from n=3.6
Γ=0.8 α = 10 cm^{-1} β = 2.0x10^{-4} mA^{-1}

I_t = (wL/β)[α+(1/L)ln(1/R)]

=10^{-5}x2.0x10^{-4}/2.0x10^{-4}[10x10^2+(1/2.0x10^{-4})ln(1/0.32)]

=67x10^{-3} =67 mA

J_t = I_t/wL = 67x10^{-3}/(10^{-5}x2.0x10^{-4}) = 3.35x10^7 Am^{-2}

For Γ=1: J_t=3.35 kAcm^{-2} and I_t=67 mA

For Γ=0.8: J_t=3.35/0.8=4.19 kAcm^{-2} and I_t=83.7 mA

Ex 6

R increased by 10% due to back-reflection from glass fibre facet. Now R_1=0.32 but R_2=0.352 so $(R_1R_2)^{1/2}$ becomes $(0.32$x$0.352)^{1/2}$ = 0.335 rather than 0.32

Now ln(1/R) = 1.094 rather than 1.139 so I_t falls by 1.041 or to 96% previous threshold.

New I_t = 67x0.96 = 64.3 mA

This will raise output power for same bias current I_b.

Ex 7

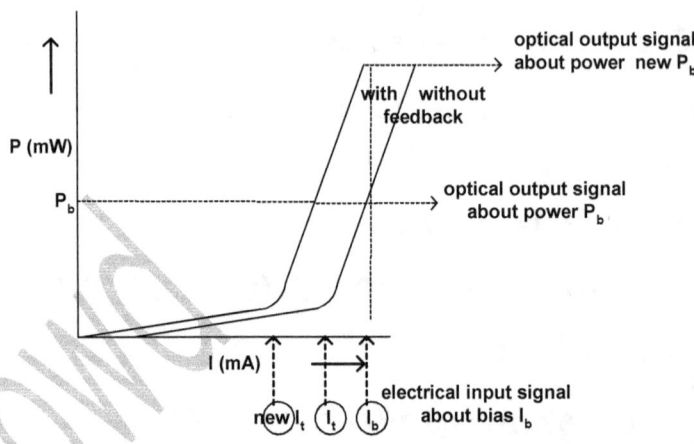

The feedback in Ex 6 has reduced I_t so at the same bias I_b the output signal is about a higher P_b as illustrated.

Note: The reverse occurs with two L-I plots for temperatures T and T' as in Figure 22 so that at the higher temperature the input signal is about a plot with *raised* threshold and "clipping" occurs on the optical output. Draw this case showing input and output modulation.

Ex 8

A typical communications laser emits 10 mW or 10dBm but a high power device is selected for 20 dBm as the link is atmospheric not fibre optic. It lases at 1.5 μm and the semiconductor alloy has average index 3.45 across the lasing spectrum. The device dimensions are stripe w=10 μm, lit depth D=1 μm, chip length L=100 μm. Calculate the photon density s.

$P = s(hc/\lambda)Vc_g$ gives $s = P\lambda/hcVc_g$

$P = 20$ dBm $= 100$ mW $= 0.1$ W

$V = 10 \times 1 \times 100 \times 10^{-18}$ m$^3 = 10^{-15}$ m^3

$c_g = 3 \times 10^8/3.45 = 0.87 \times 10^8$ ms^{-1}

$s = 0.1 \times 1.5 \times 10^{-6}/(6.6 \times 10^{-34} \times 3 \times 10^8 \times 1000 \times 10^{-18} \times 0.87 \times 10^8)$

$= 8.7 \times 10^{33}$ photons/m$^3 = 8.7 \times 10^{15}$ photons/μm^3

Sample Examination Questions

$$c = 3 \times 10^8 \quad \text{ms}^{-1}$$

$$e = 1.6 \times 10^{-19} \text{ C}$$

$$h = 6.6 \times 10^{-34} \text{ Js}$$

$$k = 1.38 \times 10^{-23} \text{ JK}^{-1}$$

$$m_c = 9.11 \times 10^{-31} \text{ kg}$$

A (a) In an injection laser the gain and "losses" are related by

$$g\frac{d}{D} = \alpha + \frac{1}{L} \ln\left(\frac{1}{R}\right).$$

Discuss the implications for diode structure taking into account the relationship between

g and J_t, the threshold current density.

(b) The reflectivity R is increased from 0.36 by 10% due to optical feedback. Calculate the
change in I_t, given the following:

$$\alpha = 12 \text{ cm}^{-1}$$

$$L = 200 \ \mu\text{m}$$

$$I_t = 50 \text{ mA (without feedback)}.$$

B For good frequency response in a LED the minority carrier lifetime τ should be small. Discuss the diode *operational* and *design* features by which this is achieved and the limitations on each approach.

In the latter case describe design for various types of confinement.

C Establish expressions for the signal-to-noise ratio for an APD and a *p-i-n* detector respectively. Consider a transmitted sinusoidal optical waveform.

Show that when large S/N ratio is required of a system it is generally operating near the shot noise limit; and that when small S/N ratio is pertinent the limit is set by circuit noise.

In the APD case show there exists an optimum avalanche gain yielding maximum S/N ratio.

D (a) Show that for a quantum well semiconductor laser the active layer thickness should satisfy

$$L_z < \left(\frac{3h^2}{8m^*kT} \right)^{\frac{1}{2}}$$

in order that only one quantised sub-level should reside within the occupied conduction band.   (Symbols have usual meanings).

(b)   What is meant by low-dimensional laser structures?   Describe examples.

E (a)   (i)   Explain your understanding of the term "internal quantum efficiency" for a LED and how to achieve close to 100% in surface emitters.

(ii)   Describe the double heterostructure in edge emitters and how this impacts on quantum efficiency.

(b)   Consider a surface emitter LED and discuss the factors that influence the recombination lifetime where high-speed response is required.   Show there is a conflict with implications
for (a) (i) above.   Show that with higher modulation current the speed improves but at a cost.

F (a)   (i)   Establish the relation for the ratio of spontaneous to stimulated emission coefficients that derives from Planck's and Boltzmann's laws.

(ii) Show that this ratio can be $\sim 10^5$ at optical fibre transmission frequencies but considerably more favourable for coherent stimulated emission at radio frequencies around 100 MHz.

(b)   Describe the effect of temperature on the threshold current for diode lasers and show that a high characteristic temperature T_0 is desirable.   Comment generally on temperature effects in this context.

G   (a)   For a photodiode in photoconductive mode, deduce the responsivity-versus-wavelength behaviour in the ideal case.   Carefully plot this result for comparison with a silicon detector explaining the divergences in performance.

(b)   Describe the gain process and related phenomena in APDs for various materials.   Discuss how the noise is quantified.

H   (a)
(i)   Explain your understanding of the term "internal quantum efficiency" for a LED and how to achieve close to 100% in surface emitters.
              25%

(ii)   Describe the double heterostructure in edge emitters and how this impacts on quantum efficiency.   25%

(b)

(i)   Consider a surface emitter LED and discuss the factors that influence the recombination lifetime where high-speed response is required.
              25%
(ii)   Show there is a conflict with implications for (a) (i) above.
         Show that with higher modulation current the speed improves but at a cost.
              25%

J   (a)
        (i)   Establish the relation for the ratio of spontaneous to stimulated emission coefficients that derives from Planck's and Boltzmann's laws.   30%
        (ii)   Show that this ratio can be $\sim 10^5$ at optical fibre transmission frequencies but considerably more favourable for coherent stimulated emission at radio frequencies around 100 MHz.
                30%

(b) Describe the effect of temperature on the threshold current for diode lasers and show that a high characteristic temperature T_0 is desirable. Comment generally on temperature effects in this context.
              40%

K (a) For a photodiode in photoconductive mode deduce the responsivity-versus-wavelength behaviour in the ideal case. Carefully plot this result for comparison with a silicon detector explaining the divergences in performance. 60%

(b) Describe the gain process and related phenomena in APDs for various materials. Discuss how the noise is quantified. 40%

Appendix 1   Periodic Table of Elements

Group:

I	II											III	IV	V	VI	VII	VIII
																	18
1																	2
H	2											13	14	15	16	17	He
1.008																	4.0026
3	4											5	6	7	8	9	10
Li	Be											B	C	N	O	F	Ne
6.94	9.0122											10.81	12.011	14.007	15.999	18.998	20.180
11	12											13	14	15	16	17	18
Na	Mg											Al	Si	P	S	Cl	Ar
22.990	24.305	3	4	5	6	7	8	9	10	11	12	26.982	28.085	30.974	32.06	35.45	39.948
19	20	21	22	23	24	25	26	27	28	29	30	31	32	33	34	35	36
K	Ca	Sc	Ti	V	Cr	Mn	Fe	Co	Ni	Cu	Zn	Ga	Ge	As	Se	Br	Kr
39.098	40.078	44.956	47.867	50.942	51.996	54.938	55.845	58.933	58.693	63.546	65.38	69.723	72.63	74.922	78.96	79.904	83.798
37	38	39	40	41	42	43	44	45	46	47	48	49	50	51	52	53	54
Rb	Sr	Y	Zr	Nb	Mo	Tc	Ru	Rh	Pd	Ag	Cd	In	Sn	Sb	Te	I	Xe
85.468	87.62	88.906	91.224	92.906	95.96	(98)	101.07	102.91	106.42	107.87	112.41	114.82	118.71	121.76	127.60	126.90	131.29
55	56	57-71	72	73	74	75	76	77	78	79	80	81	82	83	84	85	86
Cs	Ba	*	Hf	Ta	W	Re	Os	Ir	Pt	Au	Hg	Tl	Pb	Bi	Po	At	Rn
132.91	137.33		178.49	180.95	183.84	186.21	190.23	192.22	195.08	196.97	200.59	204.38	207.2	208.98	(209)	(210)	(222)
87	88	89-103	104	105	106	107	108	109	110	111	112	113	114	115	116	117	118
Fr	Ra	#	Rf	Db	Sg	Bh	Hs	Mt	Ds	Rg	Cn	Uut	Uuq	Uup	Uuh	Uus	Uuo
(223)	(226)		(265)	(268)	(271)	(270)	(277)	(276)	(281)	(280)	(285)	(284)	(289)	(288)	(293)	(294)	(294)

* Lanthanide series	57	58	59	60	61	62	63	64	65	66	67	68	69	70	71
	La	Ce	Pr	Nd	Pm	Sm	Eu	Gd	Tb	Dy	Ho	Er	Tm	Yb	Lu
	138.91	140.12	140.91	144.24	(145)	150.36	151.96	157.25	158.93	162.50	164.93	167.26	168.93	173.05	174.97
# Actinide series	89	90	91	92	93	94	95	96	97	98	99	100	101	102	103
	Ac	Th	Pa	U	Np	Pu	Am	Cm	Bk	Cf	Es	Fm	Md	No	Lr
	(227)	232.04	231.04	238.03	(237)	(244)	(243)	(247)	(247)	(251)	(252)	(257)	(258)	(259)	(262)

Groups III, IV and V have the photonic materials:

III   Al, Ga, In,

IV   Si, Ge,

V   P, As

Appendix 2   Table of Fundamental Constants

Speed light in free space   c 3×10^8 ms^{-1}

Plank's constant   h 6.6×10^{-34} Js

Electronic charge   e 1.6×10^{-19} C

Electron volt   1 eV = 1.602x10-19 J

Mass electron m_e 9.1×10^{-31} kg

Boltzmann constant   k 1.38×10^{-23} JK^{-1}

Permittivity free space ε_0 8.85×10^{-12} Fm-1

Permeability free space μ_0 1.26×10^{-6} Hm-1

Solar constant surface earth   495 Wm-2

Solar λ maximum intensity   500 nm